Student Lab Manual for Argument-Driven Inquiry in BIOLOGY

LAB INVESTIGATIONS for GRADES 9–12

Student Lab Manual for Argument-Driven Inquiry in BIOLOGY

LAB INVESTIGATIONS for GRADES 9–12

Victor Sampson, Patrick Enderle, Leeanne Gleim, Jonathon Grooms, Melanie Hester, Sherry Southerland, and Kristin Wilson

National Science Teachers Association
Arlington, Virginia

National Science Teachers Association

Claire Reinburg, Director
Wendy Rubin, Managing Editor
Amanda O'Brien, Associate Editor
Donna Yudkin, Book Acquisitions Coordinator

ART AND DESIGN
Will Thomas Jr., Director

PRINTING AND PRODUCTION
Catherine Lorrain, Director

NATIONAL SCIENCE TEACHERS ASSOCIATION
David L. Evans, Executive Director
David Beacom, Publisher

1840 Wilson Blvd., Arlington, VA 22201
www.nsta.org/store
For customer service inquiries, please call 800-277-5300.

Copyright © 2016 by the National Science Teachers Association.
All rights reserved. Printed in the United States of America.
19 18 17 16   4 3 2 1

NSTA is committed to publishing material that promotes the best in inquiry-based science education. However, conditions of actual use may vary, and the safety procedures and practices described in this book are intended to serve only as a guide. Additional precautionary measures may be required. NSTA and the authors do not warrant or represent that the procedures and practices in this book meet any safety code or standard of federal, state, or local regulations. NSTA and the authors disclaim any liability for personal injury or damage to property arising out of or relating to the use of this book, including any of the recommendations, instructions, or materials contained therein.

PERMISSIONS
Book purchasers may photocopy, print, or e-mail up to five copies of an NSTA book chapter for personal use only; this does not include display or promotional use. Elementary, middle, and high school teachers may reproduce forms, sample documents, and single NSTA book chapters needed for classroom or noncommercial, professional-development use only. E-book buyers may download files to multiple personal devices but are prohibited from posting the files to third-party servers or websites, or from passing files to non-buyers. For additional permission to photocopy or use material electronically from this NSTA Press book, please contact the Copyright Clearance Center (CCC) (*www.copyright.com*; 978-750-8400). Please access *www.nsta.org/permissions* for further information about NSTA's rights and permissions policies.

Cataloging-in-Publication Data for the e-book are also available from the Library of Congress.

LCCN: 2014000849

CONTENTS

Acknowledgments ix
About the Authors xi

SECTION 1
Introduction and Lab Safety

Introduction by Victor Sampson 3

Safety in the Science Classroom, Laboratory, or Field Sites 5

SECTION 2—Life Sciences Core Idea 1
From Molecules to Organisms: Structures and Processes

INTRODUCTION LAB

Lab 1. Osmosis and Diffusion: Why Do Red Blood Cells Appear Bigger After Being Exposed to Distilled Water?
  Lab Handout 13
  Checkout Questions 18

APPLICATION LABS

Lab 2. Cell Structure: How Should the Unknown Microscopic Organism Be Classified?
  Lab Handout 21
  Checkout Questions 25

Lab 3. Cell Cycle: Do Plant and Animal Cells Spend the Same Proportion of Time in Each Stage of the Cell Cycle?
  Lab Handout 27
  Checkout Questions 31

Lab 4. Normal and Abnormal Cell Division: Which of These Patients Could Have Cancer?
  Lab Handout 33
  Checkout Questions 38

Lab 5. Photosynthesis: Why Do Temperature and Light Intensity Affect the Rate of Photosynthesis in Plants?
  Lab Handout 40
  Checkout Questions 45

Lab 6. Cellular Respiration: How Does the Type of Food Source Affect the Rate of Cellular Respiration in Yeast?
- Lab Handout 47
- Checkout Questions 52

Lab 7. Transpiration: How Does Leaf Surface Area Affect the Movement of Water Through a Plant?
- Lab Handout 54
- Checkout Questions 59

Lab 8. Enzymes: How Do Changes in Temperature and pH Levels Affect Enzyme Activity?
- Lab Handout 61
- Checkout Questions 66

SECTION 3—Life Sciences Core Idea 2
Ecosystems: Interactions, Energy, and Dynamics

INTRODUCTION LABS

Lab 9. Population Growth: How Do Changes in the Amount and Nature of the Plant Life Available in an Ecosystem Influence Herbivore Population Growth Over Time?
- Lab Handout 71
- Checkout Questions 76

Lab 10. Predator-Prey Population Size Relationships: Which Factors Affect the Stability of a Predator-Prey Population Size Relationship?
- Lab Handout 79
- Checkout Questions 84

Lab 11. Ecosystems and Biodiversity: How Does Food Web Complexity Affect the Biodiversity of an Ecosystem?
- Lab Handout 87
- Checkout Questions 92

Lab 12. Explanations for Animal Behavior: Why Do Great White Sharks Travel Over Long Distances?
- Lab Handout 95
- Checkout Questions 101

APPLICATION LABS

Lab 13. Environmental Influences on Animal Behavior: How Has Climate Change Affected Bird Migration?
- Lab Handout 103
- Checkout Questions 108

Lab 14. Interdependence of Organisms: Why Is the Sport Fish Population of Lake Grace Decreasing in Size?
    Lab Handout 111
    Checkout Questions 122

Lab 15. Competition for Resources: How Has the Spread of the Eurasian Collared-Dove Affected Different Populations of Native Bird Species?
    Lab Handout 124
    Checkout Questions 129

SECTION 4—Life Sciences Core Idea 3
Heredity: Inheritance and Variation of Traits

INTRODUCTION LABS

Lab 16. Mendelian Genetics: Why Are the Stem and Leaf Color Traits of the Wisconsin Fast Plant Inherited in a Predictable Pattern?
    Lab Handout 135
    Checkout Questions 140

Lab 17. Chromosomes and Karyotypes: How Do Two Physically Healthy Parents Produce a Child With Down Syndrome and a Second Child With Cri Du Chat Syndrome?
    Lab Handout 142
    Checkout Questions 147

Lab 18. DNA Structure: What Is the Structure of DNA?
    Lab Handout 150
    Checkout Questions 155

Lab 19. Meiosis: How Does the Process of Meiosis Reduce the Number of Chromosomes in Reproductive Cells?
    Lab Handout 157
    Checkout Questions 162

APPLICATION LABS

Lab 20. Inheritance of Blood Type: Are All of Mr. Johnson's Children His Biological Offspring?
    Lab Handout 164
    Checkout Questions 169

Lab 21. Models of Inheritance: Which Model of Inheritance Best Explains How a Specific Trait Is Inherited in Fruit Flies?
    Lab Handout 171
    Checkout Questions 176

SECTION 5—Life Sciences Core Idea 4
Biological Evolution: Unity and Diversity

INTRODUCTION LABS

Lab 22. Biodiversity and the Fossil Record: How Has Biodiversity on Earth Changed Over Time?
    Lab Handout 181
    Checkout Questions 185

Lab 23. Mechanisms of Evolution: Why Will the Characteristics of a Bug Population Change in Different Ways in Response to Different Types of Predation?
    Lab Handout 187
    Checkout Questions 192

APPLICATION LABS

Lab 24. Descent With Modification: Does Mammalian Brain Structure Support or Refute the Theory of Descent with Modification?
    Lab Handout 194
    Checkout Questions 199

Lab 25. Mechanisms of Speciation: Why Does Geographic Isolation Lead to the Formation of a New Species?
    Lab Handout 202
    Checkout Questions 210

Lab 26. Human Evolution: How Are Humans Related to Other Members of the Family Hominidae?
    Lab Handout 212
    Checkout Questions 219

Lab 27. Whale Evolution: How Are Whales Related to Other Mammals?
    Lab Handout 221
    Checkout Questions 227

    Image Credits 229

ACKNOWLEDGMENTS

The development of this book was supported by the Institute of Education Sciences, U.S. Department of Education, through grant R305A100909 to Florida State University. The opinions expressed are those of the authors and do not represent the views of the institute or the U.S. Department of Education.

ABOUT THE AUTHORS

Victor Sampson is an associate professor of science education and the director of the Center for STEM Education at UT-Austin. He received a BA in zoology from the University of Washington, an MIT from Seattle University, and a PhD in curriculum and instruction with a specialization in science education from Arizona State University. Victor taught high school biology and chemistry for nine years before taking a position at FSU and then moving to UT-Austin. He specializes in argumentation in science education, teacher learning, and assessment. To learn more about his work in science education, go to *www.vicsampson.com*.

Patrick Enderle is an assistant professor of science education at Georgia State University. He received his BS and MS in molecular biology from East Carolina University. Patrick then spent some time as a high school biology teacher and several years as a visiting professor in the Department of Biology at East Carolina University. He then attended Florida State University (FSU), where he graduated with a PhD in science education. His research interests include argumentation in the science classroom, science teacher professional development, and enhancing undergraduate science education. To learn more about his work in science education, go to *http://patrickenderle.weebly.com*.

Leeanne Gleim received a BA in elementary education from the University of Southern Indiana and a MS in science education from FSU. While at FSU, she worked as a research assistant for Dr. Sampson. After graduating, she taught biology and honors biology at FSU Schools, where she participated in the development of the argument-driven inquiry (ADI) model. Leeanne was also responsible for writing and piloting many of the lab investigations included in this book.

Jonathon Grooms received a BS in secondary science and mathematics teaching with a focus in chemistry and physics from FSU. Upon graduation, Jonathon joined FSU's Office of Science Teaching, where he directed the physical science outreach program Science on the Move. He entered graduate school at FSU and earned a PhD in science education. Jonathon is now an assistant professor of science education in the Graduate School of Education and Human Development at The George Washington University.

Melanie Hester has a BS in biological sciences with minors in chemistry and classical civilizations from Florida State University and an MS in secondary science education from FSU. She has been teaching for more than 20 years, with the last 13 at the FSU School in Tallahassee. Melanie was a Lockheed Martin Fellow and a Woodrow Wilson fellow and received a Teacher of the Year award in 2007. She frequently gives presentations about innovative approaches to teaching at conferences and works with preservice teachers. Melanie was also responsible for writing and piloting many of the lab investigations included in this book.

ABOUT THE AUTHORS

Sherry Southerland is a professor at Florida State University and the co-director of FSU-Teach. FSU-Teach is a collaborative math and science teacher preparation program between the College of Arts and Sciences and the College of Education. She received her BS and MS in biology from Auburn University and her PhD in curriculum and instruction from Louisiana State University, with a specialization in science education and evolutionary biology. Sherry has worked as a teacher educator, biology instructor, high school science teacher, field biologist, and forensic chemist. Her research interests include understanding the influence of culture and emotions on learning—specifically evolution education and teacher education—and understanding how to better support teachers in shaping the way they approach science teaching and learning.

Kristin Wilson attended Florida State University and earned a BS in secondary science teaching with an emphasis in biology and Earth-space science. Kristin teaches biology at FSU School. She helped develop the ADI instructional model and was responsible for writing and piloting many of the lab investigations found in this book.

SECTION 1
Introduction and Lab Safety

INTRODUCTION

BY VICTOR SAMPSON

Science is much more than a body of knowledge or a set of core ideas that reflect our current understanding of how the world works and why it works that way. Science is also a set of crosscutting concepts and practices that people can use to develop and refine new explanations for, or descriptions of, the natural world. These core ideas, crosscutting concepts, and practices of science are important for you to learn. When you understand these, it is easier to appreciate the beauty and wonder of science, to engage in public discussions about science, and to critique the merits of scientific findings that are presented through the popular media. You will also have the knowledge and skills needed to continue learning about science outside school or to enter a career in science, engineering, or technology.

The core ideas of science—based on evidence from many investigations—include theories, laws, and models that scientists use to explain natural phenomena and bodies of data and to predict the outcomes of new investigations. The crosscutting concepts are themes that have value in every discipline of science and are used to help us understand a natural phenomenon. They can be used as organizational frameworks for connecting knowledge from the various fields of science into a coherent and scientifically based view of the world. Finally, the practices of science are used to develop and refine new ideas about the world. Although some practices differ from one field of science to another, all fields share a set of common practices. The practices include such things as asking and answering questions; planning and carrying out investigations; analyzing and interpreting data; and obtaining, evaluating, and communicating information. One of the most important practices of science is arguing from evidence. Arguing from evidence, or the practice of proposing, supporting, challenging, and refining claims based on evidence, is important because scientists need to be able to examine, review, and evaluate their own ideas and to critique those of others. Scientists also argue from evidence when they need to appraise the quality of data, produce and improve models, develop new testable questions from those models, and suggest ways to refine or modify existing theories, laws, and models.

It is important to always remember that science is a social activity, not an individual one. Science is social because many different scientists contribute to the development of new scientific knowledge. As scientists carry out their research, they frequently talk with their colleagues, both formally and informally. They exchange emails, engage in discussions at conferences, share research techniques and analytical procedures, and present new ideas by writing articles in journals or chapters in books. They also critique the ideas and methods used by other scientists through a formal peer review process before they can be published in journals or books. In short, scientists are members of a community, the members of which work together to build, develop, test, critique, and refine ideas. The ways

Introduction

scientists talk, write, think, and interact with each other reflect common ideas about what counts as quality and shared standards for how new ideas should be developed, shared, evaluated, and refined. These ways of interacting make science different from other ways of knowing. The core ideas, crosscutting concepts, and practices of science are important within the scientific community because most, if not all, members of that community find them to be a useful way to develop and refine new explanations for, or descriptions of, the natural world.

The laboratory investigations included in this book are designed to help you learn the core ideas, crosscutting concepts, and practices of science. During each investigation, you will have an opportunity to use a core idea, several crosscutting concepts, and the practices of science to understand a natural phenomenon or solve a problem. Your teacher will introduce each investigation by giving you a task to accomplish and a guiding question to answer. You will then work as part of a team to plan and carry out an investigation to collect the data you need to answer that question. From there, your team will develop an initial argument that includes a claim, evidence in support of your claim, and a justification of your evidence. The claim will be your answer to the guiding question, the evidence will include your analysis of the data you collected and an interpretation of that analysis, and the justification will explain why your evidence is important. Next, you will have an opportunity to share your argument with your classmates and to critique their arguments, much like professional scientists do. You will then revise your initial argument based on your colleagues' feedback. Finally, you will be asked to write an investigation report on your own to share what you learned. The report will go through double-blind peer review so you can improve it before you submit it to you teacher for a grade. As you complete more and more investigations in this lab manual, you will not only learn the core ideas associated with each investigation but also get better at using the crosscutting concepts and practices of science to understand the natural world.

SAFETY IN THE SCIENCE CLASSROOM, LABORATORY, OR FIELD SITES

Note to science teachers and supervisors/administrators:

The following safety acknowledgment form is for your use in the classroom and should be given to students at the beginning of the school year to help them understand their role in ensuring a safer and productive science experience.

Science is a process of discovering and exploring the natural world. Exploration occurs in the classroom/laboratory or in the field. As part of your science class, you will be doing many activities and investigations that will involve the use of various materials, equipment, and chemicals. Safety in the science classroom, laboratory, or field sites is the FIRST PRIORITY for students, instructors, and parents. To ensure safer classroom/laboratory/field experiences, the following **Science Rules and Regulations** have been developed for the protection and safety of all. Your instructor will provide additional rules for specific situations or settings. The rules and regulations must be followed at all times. After you have reviewed them with your instructor, read and review the rules and regulations with your parent/guardian. Their signature and your signature on the safety acknowledgment form are required before you will be permitted to participate in any activities or investigations. Your signature indicates that you have read these rules and regulations, understand them, and agree to follow them at all times while working in the classroom/laboratory or in the field.

Source: National Science Teachers Association (NSTA). Safety in the Science Classroom. *www.nsta.org/pdfs/SafetyInTheScienceClassroom.pdf.*

Safety in the Science Classroom, Laboratory, or Field Sites

Safety Standards of Student Conduct in the Classroom, Laboratory, and in the Field

1. Conduct yourself in a responsible manner at all times. Frivolous activities, mischievous behavior, throwing items, and conducting pranks are prohibited.
2. Lab and safety information and procedures must be read ahead of time. All verbal and written instructions shall be followed in carrying out the activity or investigation.
3. Eating, drinking, gum chewing, applying cosmetics, manipulating contact lenses, and other unsafe activities are not permitted in the laboratory.
4. Working in the laboratory without the instructor present is prohibited.
5. Unauthorized activities or investigations are prohibited. Unsupervised work is not permitted.
6. Entering preparation or chemical storage areas is prohibited at all times.
7. Removing chemicals or equipment from the classroom or laboratory is prohibited unless authorized by the instructor.

Personal Safety

8. Sanitized indirectly vented chemical splash goggles or safety glasses as appropriate (meeting the ANSI Z87.1 standard) shall be worn during activities or demonstrations in the classroom, laboratory, or field, including pre-laboratory work and clean-up, unless the instructor specifically states that the activity or demonstration does not require the use of eye protection.
9. When an activity requires the use of laboratory aprons, the apron shall be appropriate to the size of the student and the hazard associated with the activity or investigation. The apron shall remain tied throughout the activity or investigation.
10. All accidents, chemical spills, and injuries must be reported immediately to the instructor, no matter how trivial they may seem at the time. Follow your instructor's directions for immediate treatment.
11. Dress appropriately for laboratory work by protecting your body with clothing and shoes. This means that you should use hair ties to tie back long hair and tuck into the collar. Do not wear loose or baggy clothing or dangling jewelry on laboratory days. Acrylic nails are also a safety hazard near heat sources and should not be used. Sandals or open-toed shoes are not to be worn during any lab activities. Refer to pre-lab instructions. If in doubt, ask!

12. Know the location of all safety equipment in the room. This includes eye wash stations, the deluge shower, fire extinguishers, the fume hood, and the safety blanket. Know the location of emergency master electric and gas shut offs and exits.

13. Certain classrooms may have living organisms including plants in aquaria or other containers. Students must not handle organisms without specific instructor authorization. Wash your hands with soap and water after handling organisms and plants.

14. When an activity or investigation requires the use of laboratory gloves for hand protection, the gloves shall be appropriate for the hazard and worn throughout the activity.

Specific Safety Precautions Involving Chemicals and Lab Equipment

15. Avoid inhaling in fumes that may be generated during an activity or investigation.

16. Never fill pipettes by mouth suction. Always use the suction bulbs or pumps.

17. Do not force glass tubing into rubber stoppers. Use glycerin as a lubricant and hold the tubing with a towel as you ease the glass into the stopper.

18. Proper procedures shall be followed when using any heating or flame producing device especially gas burners. Never leave a flame unattended.

19. Remember that hot glass looks the same as cold glass. After heating, glass remains hot for a very long time. Determine if an object is hot by placing your hand close to the object but do not touch it.

20. Should a fire drill, lockdown, or other emergency occur during an investigation or activity, make sure you turn off all gas burners and electrical equipment. During an evacuation emergency, exit the room as directed. During a lockdown, move out of the line of sight from doors and windows if possible or as directed.

21. Always read the reagent bottle labels twice before you use the reagent. Be certain the chemical you use is the correct one.

22. Replace the top on any reagent bottle as soon as you have finished using it and return the reagent to the designated location.

23. Do not return unused chemicals to the reagent container. Follow the instructor's directions for the storage or disposal of these materials.

Safety in the Science Classrom, Laboratory, or Field Sites

Standards For Maintaining a Safer Laboratory Environment

24. Backpacks and books are to remain in an area designated by the instructor and shall not be brought into the laboratory area.

25. Never sit on laboratory tables.

26. Work areas should be kept clean and neat at all times. Work surfaces are to be cleaned at the end of each laboratory or activity.

27. Solid chemicals, metals, matches, filter papers, broken glass, and other materials designated by the instructor are to be deposited in the proper waste containers, not in the sink. Follow your instructor's directions for disposal of waste.

28. Sinks are to be used for the disposal of water and those solutions designated by the instructor. Other solutions must be placed in the designated waste disposal containers.

29. Glassware is to be washed with hot, soapy water and scrubbed with the appropriate type and sized brush, rinsed, dried, and returned to its original location.

30. Goggles are to be worn during the activity or investigation, clean up, and through hand washing.

31. Safety Data Sheets (SDSs) contain critical information about hazardous chemicals of which students need to be aware. Your instructor will review the salient points on the SDSs for the hazardous chemicals students will be working with and also post the SDSs in the lab for future reference.

Safety Acknowledgment Form: Science Rules and Regulations

I have read the science rules and regulations in the *Student Lab Manual for Argument-Driven Inquiry in Biology*, and I agree to follow them during any science course, investigation, or activity. By signing this form, I acknowledge that the science classroom, laboratory, or field sites can be an unsafe place to work and learn. The safety rules and regulations are developed to help prevent accidents and to ensure my own safety and the safety of my fellow students. I will follow any additional instructions given by my instructor. I understand that I may ask my instructor at any time about the rules and regulations if they are not clear to me. My failure to follow these science laboratory rules and regulations may result in disciplinary action.

_____ _____

Student Signature   Date

_____ _____

Parent/Guardian Signature   Date

SECTION 2
Life Sciences Core Idea 1:

From Molecules to Organisms: Structures and Processes

Lab 1. Osmosis and Diffusion: Why Do Red Blood Cells Appear Bigger After Being Exposed to Distilled Water?

Lab Handout

Introduction

All living things are made of cells. Some organisms, such as bacteria, are *unicellular,* which means they consist of a single cell. Other organisms, such as humans, fish, and plants, are *multicellular*, which means they consist of many cells. All cells have some parts in common. One part found in all cells is the *cell membrane*. The cell membrane surrounds the cell, holds the other parts of the cell in place, and protects the cell. Molecules such as oxygen, water, and carbon dioxide can pass in and out of the cell membrane. All cells also contain *cytoplasm*. The cytoplasm is a jelly-like substance inside the cell where most of the cell's activities take place. It's made out of water and other chemicals.

Some cells found in multicellular organisms are highly specialized and carry out very specific functions. An example of a specialized cell found in vertebrates is the erythrocyte, or red blood cell (RBC). RBCs are by far the most abundant cells in the blood. The primary function of RBCs is to transport oxygen from the lungs to the cells of the body. In the capillaries, the oxygen is released so other cells can use it. Ninety-seven percent of the oxygen that is carried by the blood from the lungs is carried by hemoglobin; the other 3% is dissolved in the plasma. Hemoglobin allows the blood to transport 30–100 times more oxygen than could be dissolved in the plasma alone.

As you can see in the figure to the right, RBCs look like little discs when they are viewed under a microscope. They have no nucleus (the nucleus is extruded from the cell as it matures to make room for more hemoglobin). A unique feature of RBCs is that they can change shape; this ability allows them to squeeze through capillaries without breaking. RBCs will also change shape in response to changes in the environment. For example, if you add a few drops of distilled water to blood on a microscope slide, the cells will look bigger after a few seconds (see the figure's right panel).

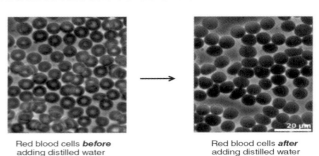

Red blood cells before and after distilled water is added

Red blood cells **before** adding distilled water

Red blood cells **after** adding distilled water

Scientists often develop and test explanations for natural phenomena. In this investigation you will have an opportunity to design and carry out an experiment to test two different explanations for why RBCs appear bigger after they are exposed to distilled water. These are the two explanations that you will test:

LAB 1

1. Molecules such as protein and polysaccharides are more concentrated inside the cell than outside the cell when the cell is in distilled water. These molecules therefore begin to move out of the cell because of the process of diffusion but are blocked by the cell membrane. As a result, these molecules push on the cell membrane and make the cell appear bigger.

2. Water molecules move into the cell because the concentration of water is greater outside the cell than it is inside the cell. As a result, water fills the cell and makes it appear bigger.

Your Task

Design and carry out an experiment to determine which of the two explanations about the appearance of RBCs after exposure to distilled water is the most valid or acceptable from a scientific perspective.

The guiding question of this investigation is, **Why do the red blood cells appear bigger after being exposed to distilled water?**

Materials

You may use any of the following materials during your investigation:

- Starch solution (starch is a polysaccharide)
- Distilled water
- Beakers
- Graduated cylinder
- Balance (electronic or triple beam)
- Dialysis tubing (assume that it behaves just like the membranes of RBCs)
- Safety goggles
- Aprons

Safety Precautions

1. Indirectly vented chemical-splash goggles and aprons are required for this activity.
2. Wash hands with soap and water after completing this lab.
3. Follow all normal lab safety rules.

Osmosis and Diffusion
Why Do Red Blood Cells Appear Bigger After Being Exposed to Distilled Water?

Getting Started

You will use models of cells rather than real cells during your experiment. You will use cell models for two reasons: (1) a model of a cell is much larger than a real cell, which makes the process of data collection much easier; and (2) you can create your cell models in any way you see fit, which makes it easier to control for a wide range of variables during your experiment. The cell models will therefore allow you to design a more informative test of the two alternative explanations outlined above.

Tying the dialysis tubing

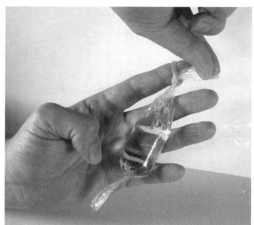

You can construct a model cell by using the dialysis tubing. Dialysis tubing behaves much like a cell membrane. To create a model of a cell, place the dialysis tubing in water until it is thoroughly soaked. Remove the soaked tubing from the water and tightly twist one end several times and either tie with string or tie a knot in the tubing. You can then fill the model cell with either a starch solution (starch is a common polysaccharide) or distilled water. Once filled, twist the open end several times and tie it tightly as shown in the figure. You can then dry the bag and place it into any type of solution you need.

In designing your experiment, you must determine what type of data you will need to collect, how you will collect it, and how you will analyze it. To determine *what type of data you will need to collect*, think about the following questions:

- What will serve as your dependent variable (e.g., mass of the cell or size of the cell)?
- What type of measurements will you need to make during your investigation?

To determine *how you will collect your data*, think about the following questions:

- What will serve as a control (or comparison) condition?
- What types of treatment conditions will you need to set up and how will you do it?
- How many "cells" will you need to use in each condition?
- How often will you collect data and when will you do it?
- How will you make sure that your data are of high quality (i.e., how will you reduce error)?
- How will you keep track of the data you collect and how will you organize the data?

To determine *how you will analyze your data*, think about the following questions:

LAB 1

- How will you determine if there is a difference between the treatment conditions and the control condition?
- What type of calculations will you need to make?
- What type of graph could you create to help make sense of your data?

Investigation Proposal Required? ☐ Yes ☐ No

Connections to Crosscutting Concepts and to the Nature of Science and the Nature of Scientific Inquiry

As you work through your investigation, be sure to think about

- the importance of identifying the underlying cause for observations,
- how models are used to study natural phenomena,
- how matter moves within or through a system,
- the difference between data and evidence in science, and
- the nature and role of experiments in science.

Argumentation Session

Once your group has finished collecting and analyzing your data, prepare a whiteboard that you can use to share your initial argument. Your whiteboard should include all the information shown in the figure below.

Argument presentation on a whiteboard

The Guiding Question:	
Our Claim:	
Our Evidence:	Our Justification of the Evidence:

To share your argument with others, we will be using a round-robin format. This means that one member of your group will stay at your lab station to share your group's argument while the other members of your group go to the other lab stations one at a time to listen to and critique the arguments developed by your classmates.

The goal of the argumentation session is not to convince others that your argument is the best one; rather, the goal is to identify errors or instances of faulty reasoning in the arguments so these mistakes can be fixed. You will therefore need to evaluate the content of the claim, the quality of the evidence used to support the claim, and the strength of the justification of the evidence included in each argument that you see. In order to critique an argument, you will need more information than what is included on the whiteboard. You might, therefore, need to ask the presenter one or more follow-up questions, such as:

- How did you collect your data? Why did you use that method? Why did you collect those data?
- What did you do to make sure the data you collected are reliable? What did you do to decrease measurement error?
- What did you do to analyze your data? Why did you decide to do it that way? Did you check your calculations?
- Is that the only way to interpret the results of your analysis? How do you know that your interpretation of your analysis is appropriate?
- Why did your group decide to present your evidence in that manner?
- What other claims did your group discuss before you decided on that one? Why did your group abandon those alternative ideas?
- How confident are you that your claim is valid? What could you do to increase your confidence?

Once the argumentation session is complete, you will have a chance to meet with your group and revise your original argument. Your group might need to gather more data or design a way to test one or more alternative claims as part of this process. Remember, your goal at this stage of the investigation is to develop the most valid or acceptable answer to the research question!

Report

Once you have completed your research, you will need to prepare an *investigation report* that consists of three sections that provide answers to the following questions:

1. What question were you trying to answer and why?
2. What did you do during your investigation and why did you conduct your investigation in this way?
3. What is your argument?

Your report should answer these questions in two pages or less. This report must be typed, and any diagrams, figures, or tables should be embedded into the document. Be sure to write in a persuasive style; you are trying to convince others that your claim is acceptable or valid!

LAB 1

Lab 1. Osmosis and Diffusion: Why Do Red Blood Cells Appear Bigger After Being Exposed to Distilled Water?

Checkout Questions

1. A model cell that contains a concentrated starch solution is placed in a beaker of distilled water. Below, complete the illustration of how this will affect the **size** of the model cell. Be sure to explain how this will affect the **mass** of the model cell.

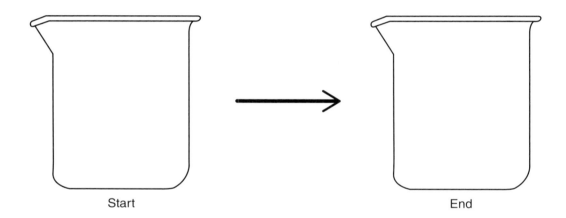

2. Observations are an example of data.

    a. I agree with this statement.
    b. I disagree with this statement.

    Explain your answer.

3. The investigation that you just completed is an example of an experiment.

    a. I agree with this statement.
    b. I disagree with this statement.

    Explain your answer, using information from your investigation about osmosis and diffusion.

LAB 1

4. Scientists often try to explain the underlying cause for their observations. Explain why this is important, using an example from your investigation about osmosis and diffusion.

5. Scientists often use models to help them understand natural phenomena. Explain what a model is and why models are important, using an example from your investigation about osmosis and diffusion.

6. Scientists often track how matter moves within or through a system they are studying. Explain why, using an example from your investigation about osmosis and diffusion.

Lab 2. Cell Structure: How Should the Unknown Microscopic Organism Be Classified?

Lab Handout

Introduction

Plant and animal cells have many organelles in common, including the nucleus, nucleolus, nuclear envelope, rough and smooth endoplasmic reticulum, Golgi apparatus, ribosomes, cell membrane, and mitochondria. Some organelles found in plant cells, however, are not found in animal cells and vice versa. For example, animal cells have centrioles (which help organize cell division in animal cells), but plant cells do not. These differences can be used to distinguish between cells that come from a plant and cells that come from an animal. The figure to the right shows animal cells from the inside of a human cheek.

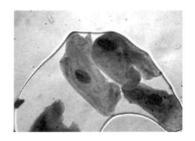

Human cheek cells

Your Task

Document the traits of an unknown microscopic organism. Then classify it based on what you know about the characteristics of plant and animal cells.

The guiding question of this investigation is, **How should the unknown microscopic organism be classified?**

Materials

You may use any of the following materials during your investigation:

- Known slide A (plant cells)
- Known slide B (plant cells)
- Known slide C (animal cells)
- Known slide D (animal cells)
- Slide with an unknown organism
- Microscope

Safety Precautions

1. Glass slides can have sharp edges—handle with care to prevent cutting of skin.

LAB 2

2. Use caution when working with electrical equipment. Keep away from water sources in that they can cause shorts, fires, and shock hazards. Use only GFI-protected circuits.
3. Wash hands with soap and water after completing this lab.
4. Follow all normal lab safety rules.

Getting Started

To answer the guiding question, you will need to conduct a systematic observation of the cell samples provided. To accomplish this task, you must first determine what type of data you will need to collect, how you will collect it, and how you will analyze it. To determine *what type of data you will need to collect*, think about the following questions:

- What type of measurements or observations will you need to make during your investigation?
- How will you quantify any differences or similarities you observe in the different cells?

To determine *how you will collect your data*, think about the following questions:

- How will you make sure that your data are of high quality (i.e., how will you reduce error)?
- How will you keep track of the data you collect and how will you organize the data?

To determine *how you will analyze your data*, think about the following questions:

- How will you define the different categories of cells (e.g., what makes a plant cell a plant cell, what makes an animal cell an animal cell)?
- What type of calculations will you need to make?
- What type of graph could you create to help make sense of your data?

Investigation Proposal Required?   ☐ Yes   ☐ No

Connections to Crosscutting Concepts and to the Nature of Science and the Nature of Scientific Inquiry

As you work through your investigation, be sure to think about

- the importance of looking for patterns during an investigation,
- how structure is related to function in organisms,
- the different type of methods that are used to answer research questions in science, and

- the importance of imagination and creativity in science.

Argumentation Session

Once your group has finished collecting and analyzing your data, prepare a whiteboard that you can use to share your initial argument. Your whiteboard should include all the information shown in the figure below.

Argument presentation on a whiteboard

The Guiding Question:	
Our Claim:	
Our Evidence:	Our Justification of the Evidence:

To share your argument with others, we will be using a round-robin format. This means that one member of your group will stay at your lab station to share your group's argument while the other members of your group go to the other lab stations one at a time to listen to and critique the arguments developed by your classmates.

The goal of the argumentation session is not to convince others that your argument is the best one; rather, the goal is to identify errors or instances of faulty reasoning in the arguments so these mistakes can be fixed. You will therefore need to evaluate the content of the claim, the quality of the evidence used to support the claim, and the strength of the justification of the evidence included in each argument that you see. In order to critique an argument, you will need more information than what is included on the whiteboard. You might, therefore, need to ask the presenter one or more follow-up questions, such as How did you collect your data? Why did you use that method? Why did you collect those data?

- What did you do to make sure the data you collected are reliable? What did you do to decrease measurement error?
- What did you do to analyze your data? Why did you decide to do it that way?
- Is that the only way to interpret the results of your analysis? How do you know that your interpretation of your analysis is appropriate?
- Why did your group decide to present your evidence in that manner?
- What other claims did your group discuss before you decided on that one? Why did your group abandon those alternative ideas?
- How confident are you that your claim is valid? What could you do to increase your confidence?

Once the argumentation session is complete, you will have a chance to meet with your group and revise your original argument. Your group might need to gather more data or design a way to test one or more alternative claims as part of this process. Remember, your

LAB 2

goal at this stage of the investigation is to develop the most valid or acceptable answer to the research question!

Report

Once you have completed your research, you will need to prepare an *investigation report* that consists of three sections that provide answers to the following questions:

1. What question were you trying to answer and why?
2. What did you do during your investigation and why did you conduct your investigation in this way?
3. What is your argument?

Your report should answer these questions in two pages or less. This report must be typed, and any diagrams, figures, or tables should be embedded into the document. Be sure to write in a persuasive style; you are trying to convince others that your claim is acceptable or valid!

Lab 2. Cell Structure: How Should the Unknown Microscopic Organism Be Classified?

Checkout Questions

1. As a biologist working for NASA you are given the task of looking for microscopic life forms on rocks brought back from Mars. Upon further investigation with a microscope, you find what you think are cells. These objects have a round shape with a defined outer boundary, a small dark-colored object in the center, and a long tail-like structure on the outside. How would you classify this object?

    a. It is an animal cell.
    b. It is a plant cell.
    c. There is not enough information to classify it.

    Why?

2. All scientists follow the same step-by-step method during an investigation.

    a. I agree with this statement.
    b. I disagree with this statement.

    Explain your answer, using information from your investigation about cell structure.

LAB 2

3. There is no room for imagination or creativity in science.

    a. I agree with this statement.
    b. I disagree with this statement.

    Explain your answer, using information from your investigation about cell structure.

4. Scientists often look for patterns in nature. Explain why this is important, using an example from your investigation about cell structure.

5. Scientists can learn a great deal about how an organism lives by looking at its structure. Explain why, using an example from your investigation about cell structure.

Lab 3. Cell Cycle: Do Plant and Animal Cells Spend the Same Proportion of Time in Each Stage of the Cell Cycle?

Lab Handout

Introduction

The cell cycle is an important process, and we need to understand it to appreciate how animals and plants are able to grow, heal, and reproduce. The figure below provides pictures of plant and animal cells in various stages of the cell cycle.

The cell cycle of (a) plant cells and (b) animal cells

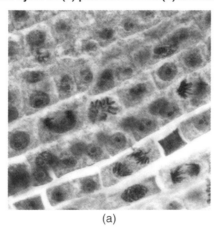
(a)

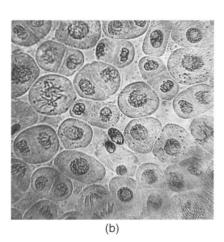
(b)

The picture of the plant cells was taken from the cells in the tip of an onion root. The roots of plants are good for studying the cell cycle because they are constantly growing and, as a result, many of the cells in the tip of the root are in the process of dividing. To create the picture in the figure (a) above, a very thin slice of onion root was placed onto a microscope slide. The root was then stained with a dye that made the chromosomes visible. These photos provide us with a clear view of the various stages of the cell cycle, yet this information tells us little about how long a cell spends in each stage and if the amount of time in each stage is different for plants and animals.

To figure out how long cells spend in each stage of the cell cycle, we need to look at the proportion of cells in a given area that are in each phase. From this information you can then determine the relative amount of time a cell spends in each stage. The portion of cells in each phase should correspond closely with the amount of time spent by each cell in each phase.

LAB 3

Your Task

Determine the proportion of time animal and plant cells spend in each phase of the cell cycle.

The guiding question of this investigation is, **Do plant and animal cells spend the same proportion of time in each stage of the cell cycle?**

Materials

You may use any of the following materials during your investigation:

- A prepared slide from an onion root tip
- A prepared slide from a whitefish blastula
- Microscope

Safety Precautions

1. Glass slides can have sharp edges—handle with care to prevent cutting of skin.
2. Use caution when working with electrical equipment. Keep away from water sources in that they can cause shorts, fires, and shock hazards. Use only GFI-protected circuits.
3. Wash hands with soap and water after completing this lab.
4. Follow all normal lab safety rules.

Getting Started

To answer the guiding question, you will need to design and conduct an investigation. You will be presented with slides that you can use to see the cells in the tip of an onion root and in a whitefish blastula. Both slides will have cells in various stages of the cell cycle. To accomplish this task, you must determine what type of data you will need to collect, how you will collect it, and how you will analyze it. To determine *what type of data you will need to collect*, think about the following questions:

- What type of measurements or observations will you need to record during your investigation?
- How will you quantify any differences or similarities you observe in the different cells?

To determine *how you will collect your data*, think about the following questions:

- How will you determine how many cells are in each stage on each slide (i.e., how many cells are in interphase, how many cells are in metaphase, and so on)?

Cell Cycle
Do Plant and Animal Cells Spend the Same Proportion of Time in Each Stage of the Cell Cycle?

- How will you make sure that your data are of high quality (i.e., how will you reduce error)?
- How will you keep track of the data you collect and how will you organize the data?

To determine how you will analyze your data, think about the following questions:

- What type of calculations will you need to make? (Hint: You will need to determine the number of cells in each stage and the total number of cells you counted and use those numbers to predict how much time a dividing cell spends in each phase.)
- What type of graph could you create to help make sense of your data?

Investigation Proposal Required?   ☐ Yes   ☐ No

Connections to Crosscutting Concepts and to the Nature of Science and the Nature of Scientific Inquiry

As you work through your investigation, be sure to think about

- the importance of identifying and explaining patterns,
- why it is important to look for proportional relationships,
- how living things move through stages of stability and change,
- the difference between observations and inferences in science, and
- how scientific knowledge can change over time.

Argumentation Session

Argument presentation on a whiteboard

The Guiding Question:	
Our Claim:	
Our Evidence:	Our Justification of the Evidence:

Once your group has finished collecting and analyzing your data, prepare a whiteboard that you can use to share your initial argument. Your whiteboard should include all the information shown in the figure to the left.

To share your argument with others, we will be using a round-robin format. This means that one member of your group will stay at your lab station to share your group's argument while the other members of your group go to the other lab stations one at a time to listen to and critique the arguments developed by your classmates.

The goal of the argumentation session is not to convince others that your argument is the best one; rather, the goal is to identify errors or instances of faulty reasoning in the arguments so these mistakes can be fixed. You will therefore need to evaluate the content of the claim, the quality of the evidence used to support the claim, and the strength of the justification of the evidence included in each argument that you see. In order to critique an argument, you will need more information than what is included on the whiteboard. You might, therefore, need to ask the presenter one or more follow-up questions, such as:

- How did you collect your data? Why did you use that method? Why did you collect those data?
- What did you do to make sure the data you collected are reliable? What did you do to decrease measurement error?
- What did you do to analyze your data? Why did you decide to do it that way? Did you check your calculations?
- Is that the only way to interpret the results of your analysis? How do you know that your interpretation of your analysis is appropriate?
- Why did your group decide to present your evidence in that manner?
- What other claims did your group discuss before you decided on that one? Why did your group abandon those alternative ideas?
- How confident are you that your claim is valid? What could you do to increase your confidence?

Once the argumentation session is complete, you will have a chance to meet with your group and revise your original argument. Your group might need to gather more data or design a way to test one or more alternative claims as part of this process. Remember, your goal at this stage of the investigation is to develop the most valid or acceptable answer to the research question!

Report

Once you have completed your research, you will need to prepare an investigation report that consists of three sections that provide answers to the following questions:

1. What question were you trying to answer and why?
2. What did you do during your investigation and why did you conduct your investigation in this way?
3. What is your argument?

Your report should answer these questions in two pages or less. This report must be typed, and any diagrams, figures, or tables should be embedded into the document. Be sure to write in a persuasive style; you are trying to convince others that your claim is acceptable or valid!

Lab 3. Cell Cycle: Do Plant and Animal Cells Spend the Same Proportion of Time in Each Stage of the Cell Cycle?

Checkout Questions

1. Describe the process of cell division.

2. Scientific knowledge never changes.

    a. I agree with this statement.
    b. I disagree with this statement.

    Explain your answer, using information from your investigation about the cell cycle.

LAB 3

3. All scientists, regardless of their background or training, will make the same observation about an event.

    a. I agree with this statement.
    b. I disagree with this statement.

    Explain your answer, using examples from your investigation about the cell cycle in plants and animals.

4. Scientists often look for patterns in nature. Explain why this is important, using an example from your investigation about the cell cycle.

5. Scientists often look for proportional relationships during an investigation. Explain why this is important, using an example from your investigation about the cell cycle.

Lab 4. Normal and Abnormal Cell Division: Which of These Patients Could Have Cancer?

Lab Handout

Introduction

Hundreds of genes control the process of cell division in normal cells. Normal cell growth requires a balance between the activity of those genes that promote cell division and those that suppress it. It also relies on the activities of genes that signal when damaged cells should undergo apoptosis (programmed cell death). Cells become cancerous after mutations accumulate in the various genes that control cell division. Some mutations occur in genes that stimulate cell division, which triggers these cells to start dividing. Other cancer-related mutations inactivate the genes that suppress cell division or those that signal the need for apoptosis. Gene mutations accumulate over time as a result of independent events.

The figure below provides an illustration of normal and cancerous cells. A normal cell often has a great deal of cytoplasm and one nucleus, and it is about the same size and shape as the cells that it borders. A cancerous cell, in contrast, often has a small amount of cytoplasm, more than one nucleus, and an abnormal shape. Cancerous cells also divide

Normal and cancer cells side by side, with normal and cancerous characteristics identified

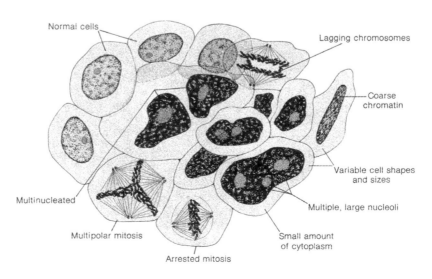

LAB 4

faster than normal cells do, so there is a greater chance that these cells will be in one of the stages of mitosis. The stages of mitosis in a cancerous cell, however, will often look different than they do in a normal cell. For example, the chromosomes may be pulled toward three or more centrioles (instead of two), and some chromosomes may lag behind others during anaphase. These types of abnormalities are often present because the genes in the cells that trigger apoptosis are no longer functional.

As a mass of cancerous cells grows, it develops into a tumor. Tumors often remain confined within the normal boundaries of a tissue during the early stages of cancer. As time passes, however, tumors will often break through the boundaries of a tissue and invade adjoining tissues. These tumors are described as malignant. Sometimes individual cancer cells will break off from a malignant tumor and travel to other parts of the body, leading to the formation of new tumors at those sites. This process is called metastasis, and it occurs during the terminal stages of cancer. Tumors that are not capable of invading adjoining tissue are described as benign.

A medical doctor will often order a procedure called a biopsy if he or she suspects that a patient has a tumor. As part of a biopsy, the doctor or other medical professional will remove a piece of tissue or a sample of cells from a patient's body so that it can be examined in a laboratory by a pathologist. The pathologist will prepare several histological slides of the tissue and use a microscope to look for the presence of cancerous cells. The pathologist will then prepare a pathology report for the medical doctor. The pathology report describes the results of the analysis and the opinion of the pathologist.

Your Task

You will be provided with images of histological slides from four different individuals. Examine these images and use what you know about the appearance of cells and what proportion of time cells tend to spend in each stage of mitosis to determine if any of the individuals have cancer.

The guiding question of this investigation is, **Which of these patients could have cancer?**

Materials

You may use any of the following materials during your investigation:

- Histological slide from stomach (thin section, H&E)
- Histological slide from pancreas (thin section, H&E)
- Images of histological slides from patients 1, 2, 3, and 4
- Microscope

Normal and Abnormal Cell Division
Which of These Patients Could Have Cancer?

Safety Precautions

1. Glass slides can have sharp edges—handle with care to prevent cutting of skin.
2. Use caution when working with electrical equipment. Keep away from water sources in that they can cause shorts, fires, and shock hazards. Use only GFI-protected circuits.
3. Wash hands with soap and water after completing this lab.
4. Follow all normal lab safety rules.

Getting Started

To answer the guiding question, you will need to design and conduct an investigation to examine the characteristics of typical cells found within the stomach and pancreas. You will then compare these cells with the cells taken from four fictitious patients. To accomplish this task, you must determine what type of data you will need to collect, how you will collect it, and how you will analyze it. To determine *what type of data you will need to collect,* think about the following questions:

- What type of measurements or observations will you need to record during your investigation? (Hint: What are the characteristics of cancerous cells?).
- Will you collect one type of data (appearance of cells only) or multiple types of data (appearance of cells and proportion of time spent in various stages of the cell cycle)?

To determine *how you will collect your data,* think about the following questions:

- What will serve as a control (or comparison) condition?
- How will you collect data? (Hint: Higher magnifications make counting cells and comparing easier.)
- How will you make sure that your data are of high quality (i.e., how will you reduce error)?
- How will you keep track of the data you collect and how will you organize the data?

To determine *how you will analyze your data,* think about the following questions:

- What type of calculations will you need to make?
- What type of graph could you create to help make sense of your data?

LAB 4

Investigation Proposal Required? ☐ Yes ☐ No

Connections to Crosscutting Concepts and to the Nature of Science and the Nature of Scientific Inquiry

As you work through your investigation, be sure to think about

- the importance of identifying patterns,
- what is and is not important at different scales or time periods,
- how structure is related to function in living things,
- the difference between observations and inferences in science, and
- the nature of scientific knowledge.

Argumentation Session

Once your group has finished collecting and analyzing your data, prepare a whiteboard that you can use to share your initial argument. Your whiteboard should include all the information shown in the figure to the right.

To share your argument with others, we will be using a round-robin format. This means that one member of your group will stay at your lab station to share your group's argument while the other members of your group go to the other lab stations one at a time to listen to and critique the arguments developed by your classmates.

Argument presentation on a whiteboard

The Guiding Question:	
Our Claim:	
Our Evidence:	Our Justification of the Evidence:

The goal of the argumentation session is not to convince others that your argument is the best one; rather, the goal is to identify errors or instances of faulty reasoning in the arguments so these mistakes can be fixed. You will therefore need to evaluate the content of the claim, the quality of the evidence used to support the claim, and the strength of the justification of the evidence included in each argument that you see. In order to critique an argument, you will need more information than what is included on the whiteboard. You might, therefore, need to ask the presenter one or more follow-up questions, such as:

- Why did you decide to focus on those data?
- What did you do to analyze your data? Why did you decide to do it that way? Did you check your calculations?
- Is that the only way to interpret the results of your analysis? How do you know that your interpretation of your analysis is appropriate?
- Why did your group decide to present your evidence in that manner?

- What other claims did your group discuss before you decided on that one? Why did your group abandon those alternative ideas?
- How confident are you that your claim is valid? What could you do to increase your confidence?

Once the argumentation session is complete, you will have a chance to meet with your group and revise your original argument. Your group might need to gather more data or design a way to test one or more alternative claims as part of this process. Remember, your goal at this stage of the investigation is to develop the most valid or acceptable answer to the research question!

Report

Once you have completed your research, you will need to prepare an investigation report that consists of three sections that provide answers to the following questions:

1. What question were you trying to answer and why?
2. What did you do during your investigation and why did you conduct your investigation in this way?
3. What is your argument?

Your report should answer these questions in two pages or less. This report must be typed, and any diagrams, figures, or tables should be embedded into the document. Be sure to write in a persuasive style; you are trying to convince others that your claim is acceptable or valid!

LAB 4

Lab 4. Normal and Abnormal Cell Division: Which of These Patients Could Have Cancer?

Checkout Questions

1. Describe cancer in terms of the cell cycle.

2. Scientists often make different inferences based on the same observations.

    a. I agree with this statement.
    b. I disagree with this statement.

    Explain your answer, using examples from your investigation about normal and abnormal cell division.

3. All scientific knowledge, including the concepts found in science textbooks, can be discarded or changed when new evidence justifies it.

    a. I agree with this statement.
    b. I disagree with this statement.

    Explain your answer, using information from your investigation about normal and abnormal cell division.

4. Scientists often look for patterns during an investigation. Explain why this is important, using an example from your investigation about normal and abnormal cell division.

5. Scientists often need to determine what is and what is not important at different scales or time periods. Explain why this is important, using an example from your investigation about normal and abnormal cell division.

6. Scientists can often tell a lot about cells by looking at their structure. Explain why, using an example from your investigation about normal and abnormal cell divisions.

LAB 5

Lab 5. Photosynthesis: Why Do Temperature and Light Intensity Affect the Rate of Photosynthesis in Plants?

Lab Handout

Introduction

You have learned that green plants have the ability to produce their own supply of sugar through the process of photosynthesis. Photosynthesis is a complex chemical process in which green plants produce sugar and oxygen for themselves. The equation for photosynthesis is as follows:

Carbon dioxide (CO_2) + water (H_2O) → sugar ($C_6H_{12}O_6$) + oxygen (O_2) + water (H_2O)

The plant uses the sugar it produces through photosynthesis to grow and produce more leaves, stems, and roots—the biomass of the plant. Plants therefore get their mass from air. The process of photosynthesis, however, does not happen all the time, and when it happens depends on a number of environmental factors. For example, plants need a supply of water, carbon dioxide, and light energy for photosynthesis to work. Plants must get these resources from the surrounding environment. The process of photosynthesis can also slow down or speed up depending on environmental conditions. In this lab investigation, you will explore how two different environmental conditions affect how quickly photosynthesis takes place within a plant. You will then develop a conceptual model that explains why.

Your Task

Design a series of experiments to determine how temperature and light intensity affect the rate of photosynthesis in spinach. Then develop a conceptual model that explains why these environmental factors affect the rate of photosynthesis in the way that they do.

The guiding question of this investigation is, **Why do temperature and light intensity affect the rate of photosynthesis in plants?**

Materials

You may use any of the following materials during your investigation:

- Spinach leaves
- Erlenmeyer flask (250 ml)
- CO_2 or O_2 gas sensor
- Sensor interface
- Beaker (600 ml or larger)

Photosynthesis
Why Do Temperature and Light Intensity Affect the Rate of Photosynthesis in Plants?

- Thermometer or temperature probe
- Hot plate
- Ring stand and clamps
- Floodlight
- Ice
- 40-W bulb
- 60-W bulb
- 100-W bulb
- Goggles and aprons

Safety Precautions

1. Safety goggles and aprons are required for this activity.
2. Use caution when working with electrical equipment. Keep away from water sources in that they can cause shorts, fires, and shock hazards. Use only GFI-protected circuits.
3. Lightbulbs and hot plates can become hot and burn skin. Handle with care!
4. Wash hands with soap and water after completing this lab.
5. Follow all normal lab safety rules.

Getting Started

The first step in developing your model is to design and carry out a series of experiments to determine how temperature and light intensity affect the rate of photosynthesis. You will therefore need a way to calculate a rate of photosynthesis. A photosynthesis rate can be calculated by measuring how much CO_2 a plant consumes or how much O_2 a plant produces over time using the following equation:

$$\text{Photosynthesis rate} = \frac{\text{change in } CO_2 \text{ or } O_2 \text{ level}}{\text{time}}$$

To measure how much CO_2 spinach consumes (or O_2 it produces) over time, simply put five spinach leaves inside a 250 ml flask and seal the flask with a CO_2 gas sensor or O_2 gas sensor. Next, fill a 600 ml (or larger) beaker with water to create a water bath in order to keep the spinach leaves at a constant temperature. You can place the flask in the water bath (see the "Equipment setup" figure on the next page) or place the water bath between the flask and the light source.

LAB 5

Equipment setup

The next step is to think about how you will collect the data and how you will analyze it. To determine *how you will collect your data*, think about the following questions:

- What will serve as a control (or comparison) condition?
- What will serve as the treatment condition(s)? (Hint: To investigate the effect of temperature on photosynthesis rate, you will need to determine how to vary the temperature inside the flask. To investigate the effect of light intensity on photosynthesis rate, you can use lightbulbs with different wattages.)
- How will you make sure that your data are of high quality (i.e., how will you reduce error)?
- How will you keep track of the data you collect and how will you organize the data?

To determine *how you will analyze your data*, think about the following questions:

- How will you determine if there is a difference between the treatment and the control conditions?
- What type of calculations will you need to make?
- What type of graph could you create to help make sense of your data?

Once you have carried out your series of experiments, your group will need to develop a conceptual model. Your model needs to explain why these two environmental factors

Photosynthesis
Why Do Temperature and Light Intensity Affect the Rate of Photosynthesis in Plants?

affect the rate of photosynthesis in the way that they do. The model should also explain what is happening at the submicroscopic level during the process of photosynthesis.

Investigation Proposal Required?   ☐ Yes   ☐ No

Connections to Crosscutting Concepts and to the Nature of Science and the Nature of Scientific Inquiry

As you work through your investigation, be sure to think about

- the importance of identifying the underlying cause for observations,
- how models are used to study natural phenomena,
- how energy and matter move within or through a system,
- the difference between observations and inferences in science, and
- the nature and role of experiments in science.

Argumentation Session

Once your group has finished collecting and analyzing your data, prepare a whiteboard that you can use to share your initial argument. Your whiteboard should include all the information shown in the figure below.

Argument presentation on a whiteboard

The Guiding Question:	
Our Claim:	
Our Evidence:	Our Justification of the Evidence:

To share your argument with others, we will be using a round-robin format. This means that one member of your group will stay at your lab station to share your group's argument while the other members of your group go to the other lab stations one at a time to listen to and critique the arguments developed by your classmates.

The goal of the argumentation session is not to convince others that your argument is the best one; rather, the goal is to identify errors or instances of faulty reasoning in the arguments so these mistakes can be fixed. You will therefore need to evaluate the content of the claim, the quality of the evidence used to support the claim, and the strength of the justification of the evidence included in each argument that you see. In order to critique an argument, you will need more information than what is included on the whiteboard. You might, therefore, need to ask the presenter one or more follow-up questions, such as:

- How did you collect your data? Why did you use that method? Why did you collect those data?

- What did you do to make sure the data you collected are reliable? What did you do to decrease measurement error?
- What did you do to analyze your data? Why did you decide to do it that way? Did you check your calculations?
- Is that the only way to interpret the results of your analysis? How do you know that your interpretation of your analysis is appropriate?
- Why did your group decide to present your evidence in that manner?
- What other claims did your group discuss before you decided on that one? Why did your group abandon those alternative ideas?
- How confident are you that your claim is valid? What could you do to increase your confidence?

Once the argumentation session is complete, you will have a chance to meet with your group and revise your original argument. Your group might need to gather more data or design a way to test one or more alternative claims as part of this process. Remember, your goal at this stage of the investigation is to develop the most valid or acceptable answer to the research question!

Report

Once you have completed your research, you will need to prepare an investigation report that consists of three sections that provide answers to the following questions:

1. What question were you trying to answer and why?
2. What did you do during your investigation and why did you conduct your investigation in this way?
3. What is your argument?

Your report should answer these questions in two pages or less. This report must be typed, and any diagrams, figures, or tables should be embedded into the document. Be sure to write in a persuasive style; you are trying to convince others that your claim is acceptable or valid!

Lab 5. Photosynthesis: Why Do Temperature and Light Intensity Affect the Rate of Photosynthesis in Plants?

Checkout Questions

1. Plant A and plant B are of the same species and are located in an environment with hot conditions. However, plant A is located in an open field and plant B is situated under a tree, receiving little sunlight. Which plant would go through the process of photosynthesis more quickly?

    a. Plant A
    b. Plant B

    Why?

2. Inferences are based on observations.

    a. I agree with this statement.
    b. I disagree with this statement.

    Explain your answer, using examples from your investigation about photosynthesis.

3. This investigation was an example of an experiment.

    a. I agree with this statement.
    b. I disagree with this statement.

    Explain your answer, using information from your investigation about photosynthesis.

LAB 5

4. An important goal in science is to explain the underlying cause for observations. Explain why this is important, using an example from your investigation about photosynthesis.

5. Scientists often use models to study or explain natural phenomenon. Explain what a model is and why models are important, using an example from your investigation about photosynthesis.

6. Scientists often need to track how matter moves in, out, and through a system during an investigation. Explain why this is important, using an example from your investigation about photosynthesis.

Lab 6. Cellular Respiration: How Does the Type of Food Source Affect the Rate of Cellular Respiration in Yeast?

Lab Handout

Introduction

One characteristic of living things is they must take in nutrients and give off waste in order to survive. This is because all living tissues (which are composed of cells) are constantly using energy. In plants, animals, and fungi this energy comes from a reaction called cellular respiration. Cellular respiration refers to a process that occurs inside cells. During this process oxygen is used to convert the chemical energy found within a molecule of sugar into a form that is usable by the organism. The following equation describes this process:

Sugar + oxygen (O_2) → water + carbon dioxide (CO_2) + adenosine triphosphate (ATP), a usable form of energy

Sugar is a generic term used to describe molecules that contain the elements carbon, hydrogen, and oxygen with the general chemical formula of $(CH_2O)n$, where n is 3 or more. Biologists also call sugars carbohydrates or saccharides. There are many different types of sugar (see the figure below). Simple sugars are called monosaccharides; examples include glucose and fructose. Complex sugars include disaccharides and polysaccharides. Examples of disaccharides include lactose, maltose, and sucrose. Examples of polysaccharides include starch, glycogen, and cellulose.

Examples of three different types of sugar

Glucose—a monosaccharide   Lactose—a disaccharide   Starch—a polysaccharide

In addition to carbohydrates there are other type of molecules found in plants and animals that could serve as potential energy sources because they also contain the elements carbon, hydrogen, and oxygen. These molecules include lipids and proteins, as

LAB 6

shown in the figure below. Lipids do not share a common molecular structure like carbohydrates. The most commonly occurring class of lipids, however, is triglycerides (fats and oils), which have a glycerol backbone bonded to three fatty acids. Proteins contain other atoms such as nitrogen and sulfur, in addition to carbon, hydrogen, and oxygen.

Examples of (a) a lipid and (b) an amino acid found in proteins

(a) Triglyceride—a lipid

(b) Lysine—an amino acid found in proteins

Yeast, like most types of fungi, produce the energy they need to survive through cellular respiration. In this investigation, you will determine if yeast can use a wide range of nutrients (e.g., proteins, fats, and different types of carbohydrates) to fuel the process of cellular respiration.

Your Task

Design a controlled experiment to determine how the type of food source available affects the rate of cellular respiration in yeast. To do this, you will need to use a CO_2 or O_2 gas sensor as shown in the figure on the next page to determine if yeast produces CO_2 (or uses O_2) at different rates in response to a change in a food source.

The guiding question of this investigation is, **How does the type of food source affect the rate of cellular respiration in yeast?**

Materials

You may use any of the following materials during your investigation:

- Yeast suspension
- Food source 1: starch (polysaccharide)
- Food source 2: sucrose (disaccharide)
- Food source 3: lactose (disaccharide)
- Food source 4: glucose (monosaccharide)
- Food source 5: protein
- Food source 6: lipid

Cellular Respiration
How Does the Type of Food Source Affect the Rate of Cellular Respiration in Yeast?

- CO_2 or O_2 gas sensor
- Sensor interface
- 2 Erlenmeyer flasks (each 250 ml)
- Ring stand and clamps
- Beaker (600 ml)
- 7 Test tubes
- Test tube rack
- Safety goggles and aprons

Safety Precautions

1. Safety goggles and aprons are required for this activity.
2. Use caution when working with electrical equipment. Keep away from water sources in that they can cause shorts, fires, and shock hazards. Use only GFI-protected circuits.
3. Wash hands with soap and water after completing this lab.
4. Follow all normal lab safety rules.

Getting Started

To answer the guiding question, you will need to design and conduct a controlled experiment. To accomplish this task, you must determine what type of data you will need to collect, how you will collect it, and how you will analyze it. To determine *what type of data you need to collect*, think about the following question:

- What type of information will you need to collect during the experiment to determine the respiration rate of yeast? (Hint: The figure to the right shows a sensor being used to measure changes in CO_2 or O_2 levels in a 250 ml flask.)

To determine *how you will collect your data*, think about the following questions:

- What will serve as the dependent variable during the experiment?
- What will serve as the independent variable?
- What other factors will you need to keep constant?
- What will serve as a control condition?

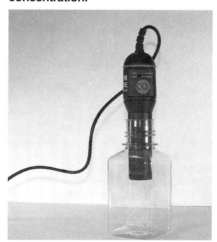

A CO_2 or an O_2 gas sensor can be used to measure changes in gas concentration.

LAB 6

- How will you make sure that your data are of high quality (i.e., how will you reduce measurement error)?
- How will you keep track of the data you collect and how will you organize the data?

To determine *how you will analyze your data*, think about the following questions:

- What type of calculations will you need to make?
- What type of graph could you create to help make sense of your data?

Investigation Proposal Required?   ☐ Yes   ☐ No

Connections to Crosscutting Concepts and to the Nature of Science and the Nature of Scientific Inquiry

As you work through your investigation, be sure to think about

- the importance of identifying the underlying cause for observations,
- how energy and matter move within or through a system,
- how structure determines function in living things,
- the difference between theories and laws in science, and
- the importance of imagination and creativity in science.

Argumentation Session

Once your group has finished collecting and analyzing your data, prepare a whiteboard that you can use to share your initial argument. Your whiteboard should include all the information shown in the figure below.

Argument presentation on a whiteboard

The Guiding Question:	
Our Claim:	
Our Evidence:	Our Justification of the Evidence:

To share your argument with others, we will be using a round-robin format. This means that one member of your group will stay at your lab station to share your group's argument while the other members of your group go to the other lab stations one at a time to listen to and critique the arguments developed by your classmates.

The goal of the argumentation session is not to convince others that your argument is the best one; rather, the goal is to identify errors or instances of faulty reasoning in the arguments so these mistakes can be fixed. You will therefore need to evaluate the content of the claim, the quality of the evidence used to support the claim, and the strength of the justification of the evidence included in each argument that you see. In order to critique an argument, you will need more information than what is included on the whiteboard. You might, therefore, need to ask the presenter one or more follow-up questions, such as:

- How did you collect your data? Why did you use that method? Why did you collect those data?
- What did you do to make sure the data you collected are reliable? What did you do to decrease measurement error?
- What did you do to analyze your data? Why did you decide to do it that way? Did you check your calculations?
- Is that the only way to interpret the results of your analysis? How do you know that your interpretation of your analysis is appropriate?
- Why did your group decide to present your evidence in that manner?
- What other claims did your group discuss before you decided on that one? Why did your group abandon those alternative ideas?
- How confident are you that your claim is valid? What could you do to increase your confidence?

Once the argumentation session is complete, you will have a chance to meet with your group and revise your original argument. Your group might need to gather more data or design a way to test one or more alternative claims as part of this process. Remember, your goal at this stage of the investigation is to develop the most valid or acceptable answer to the research question!

Report

Once you have completed your research, you will need to prepare an investigation report that consists of three sections that provide answers to the following questions:

1. What question were you trying to answer and why?
2. What did you do during your investigation and why did you conduct your investigation in this way?
3. What is your argument?

Your report should answer these questions in two pages or less. This report must be typed, and any diagrams, figures, or tables should be embedded into the document. Be sure to write in a persuasive style; you are trying to convince others that your claim is acceptable or valid!

LAB 6

Lab 6. Cellular Respiration: How Does the Type of Food Source Affect the Rate of Cellular Respiration in Yeast?

Checkout Questions

1. Describe the reactants and products of cellular respiration.

2. Scientific laws are theories that have been proven true.

    a. I agree with this statement.
    b. I disagree with this statement.

    Explain your answer, using information from your investigation about cellular respiration.

3. Scientists use their imagination and creativity when they analyze and interpret data.

    a. I agree with this statement.
    b. I disagree with this statement.

    Explain your answer, using examples from your investigation about cellular respiration.

Cellular Respiration
How Does the Type of Food Source Affect the Rate of Cellular Respiration in Yeast?

4. An important goal in science is to explain the underlying cause for observations. Explain why this is important, using an example from your investigation about cellular respiration.

5. Scientists often need to track how matter moves in, out, and through a system during an investigation. Explain why this is important, using an example from your investigation about cellular respiration.

6. Structure and function are related in living things. Explain why they are related, using an example from your investigation about cellular respiration.

LAB 7

Lab 7. Transpiration: How Does Leaf Surface Area Affect the Movement of Water Through a Plant?

Lab Handout

Introduction

Plants, just like other organisms, must be able to transport materials from one part to another. Plant transport systems consist of two large tubes made of vascular tissue that run from the roots through the shoots and to the tips of the plant. Sugars produced through the process of photosynthesis are transported through plants from leaves to roots via the vascular tissue known as the phloem. Cells use these sugars to produce the energy needed for the rest of the plant's functions. Sugars move through the plant because they are in highest concentration in the leaves, where photosynthesis takes place, and in lowest concentration in the roots. Many plants will store excess sugars in specialized root structures called tubers.

Water is transported in plants from the roots to the leaves through the vascular tissue known as the xylem. The water then enters the leaf and is used in the process of photosynthesis. In a tree such as the giant redwood of California, water must ascend over 300 ft. to reach the highest leaves. The water moves through the plant because the concentration of water is highest in the roots of a plant and lowest in the leaves. Transpiration, or loss of water from the leaves due to evaporation, helps to create a lower concentration of water (or lower osmotic potential) in the leaf. The differences in water concentration are also responsible for the movement of water from the xylem to the mesophyll layer of the leaves and subsequently out to the atmosphere (see the figure on the next page).

The transpiration rate of a plant (or how quickly water is lost from the leaves due to evaporation) is influenced by a number of environmental factors. One of the most important factors is air temperature; evaporation rates increase as the temperature goes up. Plants that live in hot locations, therefore, can lose large amounts of water from their leaves because of transpiration. When there is plenty of water in the soil, like after a heavy rain, replacing the water that is lost from the leaves because of transpiration is not a problem. However, when water is scarce and the temperature is high, plants can quickly dry out and die. Some plants, therefore, have specific adaptations that enable them to help control water loss. One such adaption could be the number or size of leaves found on a plant.

Your Task

Determine if there is a relationship between leaf surface area (i.e., the total number of leaves or the size of the leaves found on a plant) and transpiration rate.

The guiding question of this investigation is, **How does leaf surface area affect the movement of water through a plant?**

Transpiration
How Does Leaf Surface Area Affect the Movement of Water Through a Plant?

The structure of a leaf featuring the major tissues: the upper and lower epidermis, the palisade and spongy mesophyll, and the guard cells of the stoma

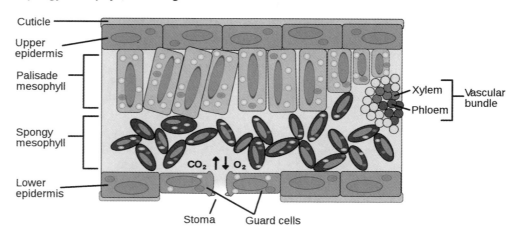

Vascular tissue (veins), made up of xylem and phloem are also shown. The light green circles within cells represent chloroplasts and indicate which tissues undergo photosynthesis.

Materials

You may use any of the following materials during your investigation:

- 6 Test tubes (150 mm × 15 mm)
- Test tube rack
- Graduated cylinder (25 ml)
- 6 Bean plants (about 3 weeks old)
- Graph paper
- Ruler
- Beaker (600 ml)
- Electronic balance
- Floodlight or plant stand with light
- Glass stirring rod
- Safety goggles and aprons

Safety Precautions

1. Safety goggles and aprons are required for this activity.
2. Use caution when working with electrical equipment. Keep away from water sources in that they can cause shorts, fires, and shock hazards. Use only GFI-protected circuits.

LAB 7

3. Lightbulbs can get hot and burn skin. Use caution and handle with care!
4. Wash hands with soap and water after completing this lab.
5. Follow all normal lab safety rules.

Plant setup

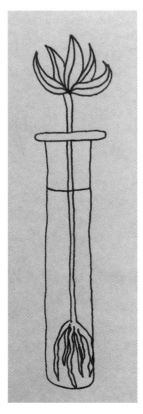

Getting Started

To answer the guiding question, you will need to design and conduct an experiment. To accomplish this task, you must be able to measure the transpiration rate of a plant. You can use the following procedure to measure a transpiration rate (see the figure to the left):

1. Pour 15 ml of tap water into a test tube.
2. Place one plant without soil on the roots into the test tube of water (be careful not to damage the roots).
3. Gently push the roots to the bottom of the tube. (The eraser end of a pencil works well for this, as does a glass stirring rod.)
4. Place this tube into a test tube rack in a warm and lighted place in the room for at least 24 hours (48 hours is better).
5. Remove the plant from the tube.
6. Remove leaves and trace on graph paper to determine total surface area.
7. Measure the amount of water left in the test tube.
8. Calculate the transpiration rate using the following equation:

$$\text{Transpiration rate} = \frac{\text{change in amount of water}}{\text{minutes}}$$

This procedure will allow you to measure the rate at which water moves through the plant. Now you must determine what type of data you will need to collect, how you will collect it, and how will you analyze it for your actual investigation. To determine *what type of data you will need to collect*, think about the following question:

- What type of measurements will you need to record during your investigation?

To determine *how you will collect your data*, think about the following questions:

- What will serve as a control (or comparison) condition?
- What types of treatment conditions will you need to set up and how will you do it?
- How many trials will you need to do?
- How often will you collect data and when will you do it?

Transpiration
How Does Leaf Surface Area Affect the Movement of Water Through a Plant?

- How will you keep track of the data you collect and how will you organize the data?

To determine *how you will analyze your data*, think about the following questions:

- How will you determine if there is a difference between the treatment conditions and the control condition?
- What type of calculations will you need to make?
- What type of graph could you create to help make sense of your data?

Investigation Proposal Required?   ☐ Yes   ☐ No

Connections to Crosscutting Concepts and to the Nature of Science and the Nature of Scientific Inquiry

As you work through your investigation, be sure to think about

- the importance of identifying the underlying cause for observations,
- what is and what is not relevant at different scales or time frames,
- how matter moves within or through a system,
- how structure determines function in living things,
- the difference between observations and inferences in science, and
- the different methods scientists can use to answer a research question in science.

Argumentation Session

Once your group has finished collecting and analyzing your data, prepare a whiteboard that you can use to share your initial argument. Your whiteboard should include all the information shown in the figure to the right.

Argument presentation on a whiteboard

The Guiding Question:	
Our Claim:	
Our Evidence:	Our Justification of the Evidence:

To share your argument with others, we will be using a round-robin format. This means that one member of your group will stay at your lab station to share your group's argument while the other members of your group go to the other lab stations one at a time to listen to and critique the arguments developed by your classmates.

The goal of the argumentation session is not to convince others that your argument is the best one; rather, the goal is to identify errors or instances of faulty reasoning in the arguments so these mistakes can be fixed. You will therefore need to evaluate the content of the claim, the quality of the evidence used to support the claim, and the strength of the justification of the evidence included in each argument that you see. In order to critique an argument, you will need more information than what is included on the whiteboard. You might, therefore, need to ask the presenter one or more follow-up questions, such as:

- How did you collect your data? Why did you use that method? Why did you collect those data?
- What did you do to make sure the data you collected are reliable? What did you do to decrease measurement error?
- What did you do to analyze your data? Why did you decide to do it that way? Did you check your calculations?
- Is that the only way to interpret the results of your analysis? How do you know that your interpretation of your analysis is appropriate?
- Why did you decide to present your evidence in that manner?
- What other claims did your group discuss before you decided on that one? Why did your group abandon those alternative ideas?
- How confident are you that your claim is valid? What could you do to increase your confidence?

Once the argumentation session is complete, you will have a chance to meet with your group and revise your original argument. Your group might need to gather more data or design a way to test one or more alternative claims as part of this process. Remember, your goal at this stage of the investigation is to develop the most valid or acceptable answer to the research question!

Report

Once you have completed your research, you will need to prepare an investigation report that consists of three sections that provide answers to the following questions:

1. What question were you trying to answer and why?
2. What did you do during your investigation and why did you conduct your investigation in this way?
3. What is your argument?

Your report should answer these questions in two pages or less. This report must be typed, and any diagrams, figures, or tables should be embedded into the document. Be sure to write in a persuasive style; you are trying to convince others that your claim is acceptable or valid!

Lab 7. Transpiration: How Does Leaf Surface Area Affect the Movement of Water Through a Plant?

Checkout Questions

1. Plant A and plant B both have the same number of leaves; however, plant B leaves are overall larger in size. If both plants were placed in a hot environment, which plant would undergo transpiration at greater rate?

    a. Plant A
    b. Plant B

    Why?

2. Observations and inferences are the same.
    a. I agree with this statement.
    b. I disagree with this statement.

    Explain your answer, using examples from your investigation about transpiration.

3. The investigation that you conducted is an example of a controlled experiment.
    a. I agree with this statement.
    b. I disagree with this statement.

    Explain your answer, using information from your investigation about transpiration.

LAB 7

4. An important goal in science is to explain the underlying cause for observations. Explain why this is important, using an example from your investigation about transpiration.

5. Scientists often need to be aware of the issue of time when designing an investigation. Explain why time frames can influence the outcomes of an investigation, using an example from your investigation about transpiration.

6. Scientists often need to track how matter moves in, out, and through a system during an investigation. Explain why this is important, using an example from your investigation about transpiration.

7. Structure is related to function in living things. Explain why, using an example from your investigation about transpiration.

Lab 8. Enzymes: How Do Changes in Temperature and pH Levels Affect Enzyme Activity?

Lab Handout

Introduction

Sugars are vital to all living organisms and are used to produce the energy (in the form of adenosine triphosphate, or ATP) an organism needs for survival. All sugars are carbohydrates, which are molecules that contain the elements carbon, hydrogen, and oxygen with the general chemical formula of $(CH_2O)n$, where n is 3 or more. Living organisms use carbohydrates as sources of energy. Different types of sugars are found in different kinds of foods, but not all of these sugars can be used as energy sources by every type of organism. In order for an organism to make

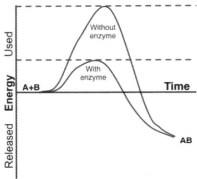

use of a sugar as an energy source, it must be capable of transporting the sugar into its cells and it must have the proper enzymes to break down the chemical bonds of the sugar to release the energy stored inside the molecule.

Enzymes are proteins that are involved in almost every chemical reaction that take place within an organism. They act as catalysts, substances that speed up chemical reactions without being destroyed or altered during the process. The figure above illustrates how an enzyme lowers the amount of energy needed for a reaction to take place, and the figure on the next page illustrates how an enzyme interacts with a substrate. Although most reactions can occur without enzymes, the rate of the reaction would be far too slow to be useful.

An example of an important enzyme in animals is catalase, which is produced in the liver and is used to catalyze the breakdown of hydrogen peroxide (H_2O_2). H_2O_2 is a toxic chemical that is produced as a natural by-product of many reactions that take place within your cells. Because it is toxic, it must be destroyed before it can do too much damage. To destroy H_2O_2, cells convert it into oxygen gas and water based on the following reaction:

$$H_2O_2 \text{ (liquid)} \xrightarrow{\text{catalase}} H_2O \text{ (liquid)} + O_2 \text{ (gas)}$$

Environmental conditions, such as temperature or pH level, can affect the function of enzymes. In this investigation, you will explore how these two environmental conditions affect enzyme activity by measuring the rate at which O_2 is produced when H_2O_2 is exposed to catalase at different pH levels and temperatures.

LAB 8

How an enzyme interacts with a substrate

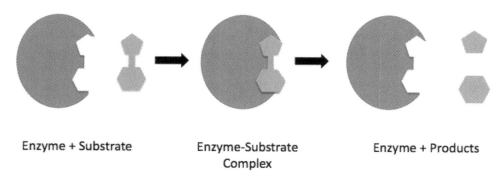

Enzyme + Substrate   Enzyme-Substrate Complex   Enzyme + Products

Your Task

Design two controlled experiments to determine how changes in temperature and pH levels affect the activity of the enzyme catalase.

The guiding question of this investigation is, **How do changes in temperature and pH levels affect enzyme activity?**

Materials

You may use any of the following materials during your investigation:

- Catalase solution
- 3% H_2O_2 solution
- 0.1 M hydrochloride (HCl) solution
- 0.1 M sodium hydroxide (NaOH) solution
- Graduated cylinder (25 ml)
- 2 Erlenmeyer flasks (each 250 ml)
- 2 Beakers (each 600 ml)
- Hot plate
- Ice
- O_2 gas sensor
- Sensor interface
- Temperature probe or thermometer
- pH probe or pH paper
- Ring stand and clamps
- Safety goggles, vinyl gloves, and aprons

Enzymes
How Do Changes in Temperature and pH Levels Affect Enzyme Activity?

Safety Precautions

1. Safety goggles, vinyl gloves, and aprons are required for this activity.
2. Use caution when working with electrical equipment. Keep away from water sources in that they can cause shorts, fires, and shock hazards. Use only GFI-protected circuits.
3. Hot plates can get hot and burn skin. Use caution and handle with care!
4. Wash hands with soap and water upon completing this lab.
5. Follow all normal lab safety rules.

Getting Started

To answer the guiding question, you will need to design and conduct two experiments. For each experiment, you must determine what type of data you will need to collect, how you will collect it, and how you will analyze it. To determine *what type of data you will need to collect*, think about the following questions:

- What will serve as your independent variable during each of your experiments?
- What will serve as your dependent variable during each of your experiments?
- What type of measurements or observations will you need to record during each of your experiments? (Hint: What information will you need to calculate a rate?)

To determine *how you will collect your data*, think about the following questions:

- What will serve as a control condition?
- What types of treatment conditions will you need to set up and how will you do it?
- How many trials will you need to conduct?
- How often will you collect data and when will you do it?
- How will you make sure that your data are of high quality (i.e., how will you reduce measurement error)?
- How will you keep track of the data you collect and how will you organize the data?

To determine *how you will analyze your data*, think about the following questions:

- How will you determine if there is a difference between the treatment conditions and the control condition?
- What type of calculations will you need to make?
- What type of graph could you create to help make sense of your data?

LAB 8

Investigation Proposal Required? ☐ Yes   ☐ No

Connections to Crosscutting Concepts and to the Nature of Science and the Nature of Scientific Inquiry.

As you work through your investigation, be sure to think about

- the importance of identifying the underlying cause for observations;
- how energy and matter move within or through a system;
- how structure is related to function in living things;
- the nature and role of experiments in science; and
- how science, as a body of knowledge, develops over time.

Argumentation Session

Once your group has finished collecting and analyzing your data, prepare a whiteboard that you can use to share your initial argument. Your whiteboard should include all the information shown in the figure to the right.

To share your argument with others, we will be using a round-robin format. This means that one member of your group will stay at your lab station to share your group's argument while the other members of your group go to the other lab stations one at a time to listen to and critique the arguments developed by your classmates.

Argument presentation on a whiteboard

The Guiding Question:	
Our Claim:	
Our Evidence:	Our Justification of the Evidence:

The goal of the argumentation session is not to convince others that your argument is the best one; rather, the goal is to identify errors or instances of faulty reasoning in the arguments so these mistakes can be fixed. You will therefore need to evaluate the content of the claim, the quality of the evidence used to support the claim, and the strength of the justification of the evidence included in each argument that you see. In order to critique an argument, you will need more information than what is included on the whiteboard. You might, therefore, need to ask the presenter one or more follow-up questions, such as:

- How did you collect your data? Why did you use that method? Why did you collect those data?
- What did you do to make sure the data you collected are reliable? What did you do to decrease measurement error?
- What did you do to analyze your data? Why did you decide to do it that way? Did you check your calculations?

Enzymes
How Do Changes in Temperature and pH Levels Affect Enzyme Activity?

- Is that the only way to interpret the results of your analysis? How do you know that your interpretation of your analysis is appropriate?
- Why did you decide to present your evidence in that manner?
- What other claims did your group discuss before you decided on that one? Why did your group abandon those alternative ideas?
- How confident are you that your claim is valid? What could you do to increase your confidence?

Once the argumentation session is complete, you will have a chance to meet with your group and revise your original argument. Your group might need to gather more data or design a way to test one or more alternative claims as part of this process. Remember, your goal at this stage of the investigation is to develop the most valid or acceptable answer to the research question!

Report

Once you have completed your research, you will need to prepare an investigation report that consists of three sections that provide answers to the following questions:

1. What question were you trying to answer and why?
2. What did you do during your investigation and why did you conduct your investigation in this way?
3. What is your argument?

Your report should answer these questions in two pages or less. This report must be typed, and any diagrams, figures, or tables should be embedded into the document. Be sure to write in a persuasive style; you are trying to convince others that your claim is acceptable or valid!

LAB 8

Lab 8. Enzymes: How Do Changes in Temperature and pH Levels Affect Enzyme Activity?

Checkout Questions

1. How do environmental factors, such as temperature and pH, affect enzyme function?

2. All investigations are experiments.

    a. I agree with this statement.
    b. I disagree with this statement.

    Explain your answer, using information from your investigation about enzymes.

3. Scientific knowledge that is based on a well-designed experiment will not change.

    a. I agree with this statement.
    b. I disagree with this statement.

    Explain your answer, using examples from your investigation about enzymes.

Enzymes
How Do Changes in Temperature and pH Levels Affect Enzyme Activity?

4. An important goal in science is to explain the underlying cause for observations. Explain why this is important, using an example from your investigation about enzymes.

5. Scientists often need to track how matter moves in, out, and through a system during an investigation. Explain why this is important, using an example from your investigation about enzymes.

6. Structure and function are related in living things. Explain why, using an example from your investigation about enzymes.

SECTION 3
Life Sciences Core Idea 2:

Ecosystems: Interactions, Energy, and Dynamics

Lab 9. Population Growth: How Do Changes in the Amount and Nature of the Plant Life Available in an Ecosystem Influence Herbivore Population Growth Over Time?

Lab Handout

Introduction

A population is a group of individuals that belong to the same species and live in the same region at the same time (the figure to the right shows an example of a rabbit population). Populations have unique attributes such as growth rate, age structure, sex ratio, birth rate, and death rate. The growth rate of population describes how the size of the population changes over a set time period. The age structure refers to the distribution of individuals based on age. The sex ratio is the proportion of males and females in the population. The birth rate is the frequency of births within a population over a set time period. The death rate is the frequency of deaths over a set time period. The characteristics of a population can change over time because of births, deaths, and the dispersal of individuals from one population to another.

The population of rabbits at the Myxomatosis Trial Enclosure on Wardang Island, Australia

Populations of animals interact with each other and their environment in a variety of ways. One of the primary ways a population interacts with the environment and with other populations is through feeding. Animals can eat plants, other animals, or both. Animals that feed on plants are called herbivores. The plants that herbivores eat, however, are not all the same. Some plants grow quickly and are plentiful, which makes them easy to find, whereas others grow slowly and are sparse. Some plants are drought resistant, whereas others do not grow well unless there is plenty of water available. Finally, and perhaps most important, some plants are loaded with nutrients (vitamins and minerals) but low in calories, some are high in calories but have fewer nutrients, and some are high in both calories and nutrients.

There are a number of factors that might influence the size of a herbivore population in an ecosystem. These factors include, but are not limited to, the amount of food available to eat, the type of plants available to eat, and the nutritional value of these plants. In this investigation, you will explore how the size of a herbivore population changes over time in response to changes in the nature and type of plants available for it to eat.

LAB 9

Your Task

Explain how the size of a population of rabbits (herbivores) changes over time in response to changes in the amounts and characteristics of the plants available in an ecosystem.

The guiding question of this investigation is, **How do changes in the amount and nature of the plant life available in an ecosystem influence herbivore population growth over time?**

Materials

You will use an online simulation called *Rabbits Grass Weeds* to conduct your investigation. You can access the simulation by going to the following website: *http://ccl.northwestern.edu/netlogo/models/RabbitsGrassWeeds*.

Safety Precautions

1. Use caution when working with electrical equipment. Keep away from water sources in that they can cause shorts, fires, and shock hazards. Use only GFI-protected circuits.

2. Wash hands with soap and water after completing this lab.

3. Follow all normal lab safety rules.

Getting Started

The *Rabbits Grass Weeds* simulation allows you to explore a simple ecosystem made up of rabbits, grass, and weeds (see the figure on the next page). The rabbits wander around randomly, and the grass and weeds grow randomly. When a rabbit bumps into some grass or weeds, it eats the grass and gains energy. If the rabbit gains enough energy, it reproduces. If it doesn't gain enough energy, it dies. The grass and weeds can be adjusted to grow at different rates and give the rabbits differing amounts of energy.

This simulation is easy to use. Click the SETUP button to set up the ecosystem with rabbits and grass, then click the GO button to start the simulation. It is also easy to adjust the characteristics of the simulated ecosystem. The NUMBER slider controls the initial number of rabbits (0–500). The BIRTH-THRESHOLD slider sets the energy level at which the rabbits reproduce (0–20). Rabbits can reproduce at any time when the threshold is set at zero. When the threshold is set at 20, a rabbit must eat enough food to have an energy level of 20 before it can reproduce. The GRASS-GROWTH-RATE slider controls the rate at which the grass grows (0–20). When the grass growth rate is set to 0, no grass will grow in the simulated ecosystem. The WEEDS-GROWTH-RATE slider controls the rate at which the weeds grow (0–20). The GRASS-ENERGY slider and the WEED-ENERGY slider allow you to set the amount of energy a rabbit can get from a plant when it is eaten (0–10).

Population Growth

How Do Changes in the Amount and Nature of the Plant Life Available in an Ecosystem Influence Herbivore Population Growth Over Time?

To answer the guiding question, you must determine what type of data you will need to collect, how you will collect it, and how you will analyze it. To determine *what type of data you will need to collect*, think about the following questions:

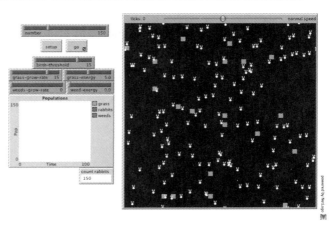

A screen shot from the *Rabbits Grass Weeds* simulation

- What will serve as your independent variables (presence of grass, presence of weeds, grass growth rate, amount of energy obtained from grass, weed growth rate, and so on)?
- What will serve as your dependent variable (population size of rabbits, population size of weeds, population size of grass, and so on)?
- What type of measurements or observations will you need to record during your investigation?

To determine *how you will collect your data*, think about the following questions:

- What will serve as a control (or comparison) condition?
- What types of treatment conditions will you need to set up and how will you do it?
- How long will you need to run each simulation?
- How often will you collect data and when will you do it?
- How will you make sure that your data are of high quality (i.e., how will you reduce measurement error)?
- How will you keep track of the data you collect and how will you organize the data?

To determine *how you will analyze your data*, think about the following questions:

- How will you determine if there is a difference between the treatment conditions and the control condition?
- What type of calculations will you need to make?
- What type of graph could you create to help make sense of your data?

LAB 9

Investigation Proposal Required? ☐ Yes   ☐ No

Connections to Crosscutting Concepts and to the Nature of Science and the Nature of Scientific Inquiry

As you work through your investigation, be sure to think about

- the importance of identifying patterns,
- how models are used to study natural phenomena,
- how living things or systems go through periods of stability and change,
- the different types of investigations that can be designed and carried out by scientists, and
- the difference between data and evidence in science.

Argumentation Session

Once your group has finished collecting and analyzing your data, prepare a whiteboard that you can use to share your initial argument. Your whiteboard should include all the information shown in the figure below.

Argument presentation on a whiteboard

The Guiding Question:	
Our Claim:	
Our Evidence:	Our Justification of the Evidence:

To share your argument with others, we will be using a round-robin format. This means that one member of your group will stay at your lab station to share your group's argument while the other members of your group go to the other lab stations one at a time to listen to and critique the arguments developed by your classmates.

The goal of the argumentation session is not to convince others that your argument is the best one; rather, the goal is to identify errors or instances of faulty reasoning in the arguments so these mistakes can be fixed. You will therefore need to evaluate the content of the claim, the quality of the evidence used to support the claim, and the strength of the justification of the evidence included in each argument that you see. In order to critique an argument, you will need more information than what is included on the whiteboard. You might, therefore, need to ask the presenter one or more follow-up questions, such as:

- How did you use the simulation to collect your data?
- What did you do to analyze your data? Why did you decide to do it that way? Did you check your calculations?

Population Growth
How Do Changes in the Amount and Nature of the Plant Life Available in an Ecosystem Influence Herbivore Population Growth Over Time?

- Is that the only way to interpret the results of your analysis? How do you know that your interpretation of your analysis is appropriate?
- Why did you decide to present your evidence in that manner?
- What other claims did your group discuss before you decided on that one? Why did your group abandon those alternative ideas?
- How confident are you that your claim is valid? What could you do to increase your confidence?

Once the argumentation session is complete, you will have a chance to meet with your group and revise your original argument. Your group might need to gather more data or design a way to test one or more alternative claims as part of this process. Remember, your goal at this stage of the investigation is to develop the most valid or acceptable answer to the research question!

Report

Once you have completed your research, you will need to prepare an investigation report that consists of three sections that provide answers to the following questions:

1. What question were you trying to answer and why?
2. What did you do during your investigation and why did you conduct your investigation in this way?
3. What is your argument?

Your report should answer these questions in two pages or less. This report must be typed, and any diagrams, figures, or tables should be embedded into the document. Be sure to write in a persuasive style; you are trying to convince others that your claim is acceptable or valid!

LAB 9

Lab 9. Population Growth: How Do Changes in the Amount and Nature of the Plant Life Available in an Ecosystem Influence Herbivore Population Growth Over Time?

Checkout Questions

1. In the space below, draw two graphs: the one on the left should show how the size of a population would change over time with unlimited natural resources (food, water, space), and the one on the right should show how the size of a population would grow over time with a limited amount of natural resources.

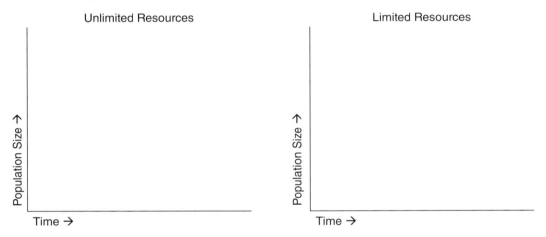

Explain your reasoning. Why did you draw your graphs in the way you did?

Population Growth
How Do Changes in the Amount and Nature of the Plant Life Available in an Ecosystem Influence Herbivore Population Growth Over Time?

2. The method used by a scientist during an investigation depends on what is being studied and the nature of the research question he or she is trying to answer.

    a. I agree with this statement.
    b. I disagree with this statement.

    Explain your answer, using examples from your investigation about population growth.

3. The term *data* and the term *evidence* have the same meaning in science.

    a. I agree with this statement.
    b. I disagree with this statement.

    Explain your answer, using information from your investigation about population growth.

4. Scientists often look for patterns in nature. Explain why this is important, using an example from your investigation about population growth.

LAB 9

5. Scientists often use models to study complex natural phenomenon. Explain what a model is and why models are valuable in science, using an example from your investigation about population growth.

6. Biological systems, such as ecosystems, often go through periods of stability and change. Explain what this means, using an example from your investigation about population growth.

Lab 10. Predator-Prey Population Size Relationships: Which Factors Affect the Stability of a Predator-Prey Population Size Relationship?

Lab Handout

Introduction

Several factors determine the size of any population within an ecosystem. The factors that affect the size of a population are divided into two broad categories: abiotic factors, which are the nonliving components of an ecosystem, and biotic factors, which are the other living components found within an ecosystem.

Wolves (predators) surrounding a bison (prey)

Predation is an example of a biotic factor that influences the size of a population (see the figure to the right). Predation is an interaction between species in which one species (the predator) uses another species as food (the prey). Predation often leads to an increase in the population size of the predator and a decrease in the population size of the prey. However, if the size of a prey population gets too small, many of the predators may not have enough food to eat and will die. As a result, the predator population size and the population size of its prey are linked. The sizes of a predator population and a prey population often cycle over several generations (see the figure to the right, "A stable predator-prey population size relationship"), and this cyclic pattern is often described as a predator-prey population size relationship. A predator-prey population size relationship that results in both populations surviving over time, despite fluctuations in the size of each one over several generations, is described as stable. A predator-prey relationship that results in the extinction of one or more species, in contrast, is described as unstable.

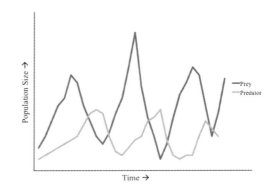

A stable predator-prey population size relationship

There are a number of factors that might influence the size of predator and prey populations in an ecosystem and can contribute to the overall stability of a predator-prey population size relationship. These factors include, but are not limited to, the amount of food available for the prey, the number of different prey species available for a predator, and how fast the predator and the prey species reproduce. In this investigation, you will investigate how a population of predators (wolves) and a population of its prey (sheep) interact with each other and the plant life in an environment over time.

Your Task

Determine what makes a predator-prey population size relationship stable or unstable.

The guiding question of this investigation is, **Which factors affect the stability of a predator-prey population size relationship?**

Materials

You will use an online simulation called *Wolf Sheep Predation* to conduct your investigation. You can access the simulation by going to the following website: *http://ccl.northwestern.edu/netlogo/models/WolfSheepPredation*.

Safety Precautions

1. Use caution when working with electrical equipment. Keep away from water sources in that they can cause shorts, fires, and shock hazards. Use only GFI-protected circuits.
2. Wash hands with soap and water after completing this lab.
3. Follow all normal lab safety rules.

Getting Started

The *Wolf Sheep Predation* simulation allows you to explore the stability of the predator-prey population size relationship (see the figure at the top of the next page) between a population of wolves (the predator) and a population of sheep (the prey). In the simulation, wolves and sheep wander around the landscape at random. The wolves lose energy with each step, and when they run out of energy they die. The wolves therefore must eat sheep to replenish their energy. You can set the simulation so there is an unlimited amount of food for the sheep to eat (grass off) or you can set the simulation so it includes a limited amount of grass in the ecosystem (grass on). If you decide to leave grass out of the simulation, the sheep never run out of energy and they only die when a wolf eats them. If you decide to include grass in the simulation, the sheep must eat grass to maintain their energy; when they run out of energy, they die. Once grass is eaten by a sheep it will only regrow after a fixed amount of time; you can adjust the amount of time it takes for grass to regrow. You can also set other factors such as the initial population size of the wolves and the sheep

Predator-Prey Population Size Relationships
Which Factors Affect the Stability of a Predator-Prey Population Size Relationship?

and what percentage of the wolves and sheep reproduce with each "tick" of the simulation (each tick represents a set amount of time—in this case a day).

To answer the guiding question, you must determine what type of data you will need to collect, how you will collect it, and how you will analyze it. To determine *what type of data you will need to collect*, think about the following questions:

A screen shot from the *Wolf Sheep Predation* simulation

- How will you determine if a predator-prey relationship is stable?
- What will serve as your dependent variable (number of wolves, number of sheep, and so on)?
- What type of measurements or observations will you need to record during your investigation?

To determine *how you will collect your data*, think about the following questions:

- What will serve as a control (or comparison) condition?
- What types of treatment conditions will you need to set up and how will you do it?
- How often will you collect data and when will you do it?
- How will you make sure that your data are of high quality (i.e., how will you reduce error)?
- How will you keep track of the data you collect and how will you organize the data?

To determine *how you will analyze your data*, think about the following questions:

- How will you determine if there is a difference between the treatment conditions and the control condition?
- What type of calculations will you need to make?
- What type of graph could you create to help make sense of your data?

LAB 10

Investigation Proposal Required? ☐ Yes ☐ No

Connections to Crosscutting Concepts and to the Nature of Science and the Nature of Scientific Inquiry

As you work through your investigation, be sure to think about

- the importance of identifying patterns,
- the importance of identifying the underlying cause for observations,
- how models are used to study natural phenomena,
- how systems go through periods of stability and change,
- how social and cultural factors influence the work of scientists, and
- different methods used in scientific investigations.

Argumentation Session

Once your group has finished collecting and analyzing your data, prepare a whiteboard that you can use to share your initial argument. Your whiteboard should include all the information shown in the figure below.

Argument presentation on a whiteboard

The Guiding Question:	
Our Claim:	
Our Evidence:	Our Justification of the Evidence:

To share your argument with others, we will be using a round-robin format. This means that one member of your group will stay at your lab station to share your group's argument while the other members of your group go to the other lab stations one at a time to listen to and critique the arguments developed by your classmates.

The goal of the argumentation session is not to convince others that your argument is the best one; rather, the goal is to identify errors or instances of faulty reasoning in the arguments so these mistakes can be fixed. You will therefore need to evaluate the content of the claim, the quality of the evidence used to support the claim, and the strength of the justification of the evidence included in each argument that you see. In order to critique an argument, you will need more information than what is included on the whiteboard. You might, therefore, need to ask the presenter one or more follow-up questions, such as:

- How did you use the simulation to collect your data?
- What did you do to analyze your data? Why did you decide to do it that way? Did you check your calculations?
- Is that the only way to interpret the results of your analysis? How do you know that your interpretation of your analysis is appropriate?

- Why did your group decide to present your evidence in that manner?
- What other claims did your group discuss before you decided on that one? Why did your group abandon those alternative ideas?
- How confident are you that your claim is valid? What could you do to increase your confidence?

Once the argumentation session is complete, you will have a chance to meet with your group and revise your original argument. Your group might need to gather more data or design a way to test one or more alternative claims as part of this process. Remember, your goal at this stage of the investigation is to develop the most valid or acceptable answer to the research question!

Report

Once you have completed your research, you will need to prepare an investigation report that consists of three sections that provide answers to the following questions:

1. What question were you trying to answer and why?
2. What did you do during your investigation and why did you conduct your investigation in this way?
3. What is your argument?

Your report should answer these questions in two pages or less. This report must be typed, and any diagrams, figures, or tables should be embedded into the document. Be sure to write in a persuasive style; you are trying to convince others that your claim is acceptable or valid!

LAB 10

Lab 10. Predator-Prey Population Size Relationships: Which Factors Affect the Stability of a Predator-Prey Population Size Relationship?

Checkout Questions

1. In the space below, draw two graphs: the one on the left should show how the size of a predator population and the size of a prey population fluctuate over time when the predator-prey population size relationship is stable; the one on the right should show how the size of a predator population and the size of a prey population fluctuate over time when the predator-prey population size relationship is unstable.

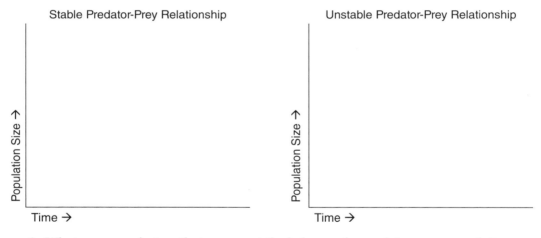

2. What are some factors that can upset the balance of a predator-prey population size relationship, and why do these factors make these relationships unstable?

Predator-Prey Population Size Relationships
Which Factors Affect the Stability of a Predator-Prey Population Size Relationship?

3. Social and cultural values or expectations influence how an investigation is designed and the extent to which the findings are acceptable.

    a. I agree with this statement.
    b. I disagree with this statement.

    Explain your answer, using information from your investigation about predator-prey population size relationships.

4. Scientists reach true and accurate conclusions when they use the scientific method.

    a. I agree with this statement.
    b. I disagree with this statement.

    Explain your answer, using examples from your investigation about predator-prey population size relationships.

LAB 10

5. Scientists often look for patterns in nature. Explain why patterns are important, using an example from your investigation about predator-prey population size relationships.

6. A major goal of scientists is to identify the underlying cause for natural phenomena. Explain why it is important to learn about underlying causes, using an example from your investigation about predator-prey population size relationships.

7. Scientists often use models to study complex natural phenomenon. Explain what a model is and why models are valuable in science, using an example from your investigation about predator-prey population size relationships.

Lab 11. Ecosystems and Biodiversity: How Does Food Web Complexity Affect the Biodiversity of an Ecosystem?

Lab Handout

Introduction

An ecosystem is a community of living organisms and the nonliving components of the environment. Energy flows in an ecosystem in one direction through food chains, and a food web is made up of all the food chains within a community of organisms. Food chains and food webs consist of the producers (the autotrophs of an ecosystem), the primary consumers (the herbivores and omnivores of the ecosystem), the secondary consumers (the carnivores and omnivores of the ecosystem), and the top predator. Some ecosystems have complex food webs and some do not. In ecosystems with a complex food web, herbivores and omnivores eat many different types of plants and the carnivores eat many different types of animals. The consumers in this type of ecosystem are described as generalists. Ecosystems that support consumers that rely on a single food source, in contrast, have simple food webs, because the consumers are specialists. An example of a complex food web is provided in panel (a) of the figure on the next page, and an example of a simple food web is provided in panel (b) of that figure.

Biodiversity refers to the variation in species found within an ecosystem, and it is measured in two ways: (1) species richness, which is the total number of different species in an ecosystem; and (2) relative abundance, which is a measure of how common each species is within the ecosystem. Regions that are home to many different species with a high relative abundance of those different species have high levels of biodiversity, whereas regions with only a few different types of species or that have moderate species richness but a low relative abundance of several species have a low level of biodiversity.

Notice that the food webs illustrated on the next page have the same amount of species richness even though the feeding relationships are different. Some of the feeding relationships illustrated in these two ecosystems, however, may or may not be sustainable over time and may result in a net decrease in biodiversity. The relative abundance of each species, for example, may change if one or more of the populations within the ecosystem grows or declines over time. The species richness of the ecosystems could also change if some of the populations disappear because of too much predation or too little access to natural resources. Given the role that biodiversity plays in ecosystem health and tolerance to ecological disturbances, it is important to understand how food web complexity is related to the biodiversity of an ecosystem.

LAB 11

Example of (a) a complex food web and (b) a simple food web

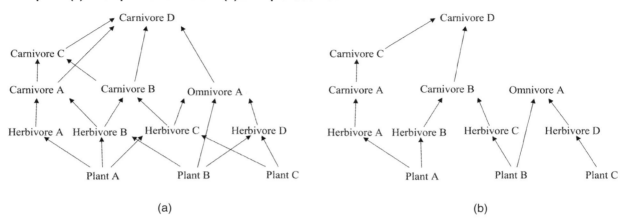

Your Task

Use the online simulation *Ecology Lab* (see the figure on p. 89), to explore the relationship between food web complexity and biodiversity in an ecosystem.

The guiding question of this investigation is, **How does food web complexity affect the biodiversity of an ecosystem?**

Materials

You will use an online simulation called *Ecology Lab* to conduct your investigation. You can access the simulation by going to the following website: *www.learner.org/courses/envsci/interactives/ecology*.

Safety Precautions

1. Use caution when working with electrical equipment. Keep away from water sources in that they can cause shorts, fires, and shock hazards. Use only GFI-protected circuits.

2. Wash hands with soap and water after completing this lab.

3. Follow all normal lab safety rules.

Getting Started

The *Ecology Lab* simulation allows you to create different food chains and webs within a model ecosystem. Once you establish the food chains and webs in the model ecosystem, you can run the simulation to determine the effect on the population of each organism (see the figure on p. 89).

Ecosystems and Biodiversity
How Does Food Web Complexity Affect the Biodiversity of an Ecosystem?

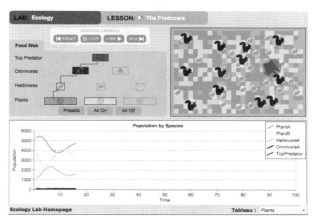

A screen shot of the *Ecology Lab* simulation

To answer the guiding question, you will need to design and conduct several experiments using the online simulation. To accomplish this task, you must determine what type of data you will need to collect during each experiment, how you will collect it, and how you will analyze it.

To determine *what type of data you will need to collect,* think about the following questions:

- What will serve as your dependent variable (population size, number of different populations, relative abundance, and so on)?
- What type of data will you need to keep a record of during your investigation?

To determine *how you will collect your data,* think about the following questions:

- What will serve as a control (or comparison) condition during each experiment?
- What types of treatment conditions will you need to set up for each experiment?
- What variables will you need to control during each experiment?
- How often will you collect data and when will you do it?
- How will you keep track of the data you collect and how will you organize the data?

To determine *how you will analyze your data,* think about the following questions:

- How will you determine if there is a difference between the conditions during each experiment?
- What type of calculations will you need to make?
- What type of table or graph could you create to help make sense of your data?

Investigation Proposal Required?   ☐ Yes   ☐ No

Connections to Crosscutting Concepts and to the Nature of Science and the Nature of Scientific Inquiry

As you work through your investigation, be sure to think about

- the importance of identifying patterns,

Student Lab Manual for Argument-Driven Inquiry in Biology

- the importance of identifying the underlying cause for observations,
- how models are used to study natural phenomena,
- how systems go through periods of stability and change,
- the importance of creativity and imagination in science, and
- the factors that influence observations and inferences in science.

Argumentation Session

Once your group has finished collecting and analyzing your data, prepare a whiteboard that you can use to share your initial argument. Your whiteboard should include all the information shown in the figure to the right.

To share your argument with others, we will be using a round-robin format. This means that one member of your group will stay at your lab station to share your group's argument while the other members of your group go to the other lab stations one at a time to listen to and critique the arguments developed by your classmates.

Argument presentation on a whiteboard

The Guiding Question:	
Our Claim:	
Our Evidence:	Our Justification of the Evidence:

The goal of the argumentation session is not to convince others that your argument is the best one; rather, the goal is to identify errors or instances of faulty reasoning in the arguments so these mistakes can be fixed. You will therefore need to evaluate the content of the claim, the quality of the evidence used to support the claim, and the strength of the justification of the evidence included in each argument that you see. In order to critique an argument, you will need more information than what is included on the whiteboard. You might, therefore, need to ask the presenter one or more follow-up questions such as:

- How did you use the simulation to collect your data?
- What did you do to analyze your data? Why did you decide to do it that way? Did you check your calculations?
- Is that the only way to interpret the results of your analysis? How do you know that your interpretation of your analysis is appropriate?
- Why did your group decide to present your evidence in that manner?
- What other claims did your group discuss before you decided on that one? Why did your group abandon those alternative ideas?
- How confident are you that your claim is valid? What could you do to increase your confidence?

Once the argumentation session is complete, you will have a chance to meet with your group and revise your original argument. Your group might need to gather more data or design a way to test one or more alternative claims as part of this process. Remember, your goal at this stage of the investigation is to develop the most valid or acceptable answer to the research question!

Report

Once you have completed your research, you will need to prepare an investigation report that consists of three sections that provide answers to the following questions:

1. What question were you trying to answer and why?
2. What did you do during your investigation and why did you conduct your investigation in this way?
3. What is your argument?

Your report should answer these questions in two pages or less. This report must be typed, and any diagrams, figures, or tables should be embedded into the document. Be sure to write in a persuasive style; you are trying to convince others that your claim is acceptable or valid!

LAB 11

Lab 11. Ecosystems and Biodiversity: How Does Food Web Complexity Affect the Biodiversity of an Ecosystem?

Checkout Questions

Use the figure below to answer questions 1 and 2. The figure illustrates the food webs of two different ecosystems. Ecosystem A has a simple food web and Ecosystem B has a complex one.

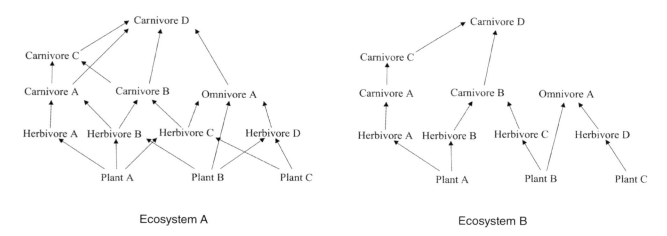

Ecosystem A   Ecosystem B

1. Which ecosystem has greater biodiversity?

    a. Ecosystem A
    b. Ecosystem B
    c. Ecosystems A and B have the same amount of biodiversity
    d. Unable to determine from the information provided

   Explain your answer.

Ecosystems and Biodiversity
How Does Food Web Complexity Affect the Biodiversity of an Ecosystem?

2. Which ecosystem is most likely to sustain a greater amount of biodiversity over time?

    a. Ecosystem A
    b. Ecosystem B
    c. Ecosystems A and B will both sustain a large amount of biodiversity
    d. Unable to determine from the information provided

    Explain your answer.

3. The inferences that are made by a scientist are influenced by his or her background and past experiences, but the observations made by a scientist are not.

    a. I agree with this statement.
    b. I disagree with this statement.

    Explain your answer, using examples from your investigation about ecosystems.

4. Science requires logic and reason but not imagination or creativity.

    a. I agree with this statement.
    b. I disagree with this statement.

    Explain your answer, using information from your investigation about biodiversity.

LAB 11

5. Scientists often attempt to identify patterns in nature. Explain why the identification of patterns is useful in science, using an example from your investigation about ecosystems.

6. An important goal in science is to identify the underlying cause of a natural phenomenon. Explain why it is important for scientists to learn about underlying causes, using an example from your investigation about ecosystems.

7. Scientists often use models to study complex natural phenomenon. Explain what a model is and how you used models during your investigation about ecosystems.

8. Biological systems, such as ecosystems, often go through periods of stability and change. Explain what this means, using an example from your investigation about ecosystems.

Lab 12. Explanations for Animal Behavior: Why Do Great White Sharks Travel Over Long Distances?

Lab Handout

Introduction

Shark populations worldwide are declining in areas where they were once common. As a result, the International Union for Conservation of Nature (IUCN), has classified many species of shark as threatened with extinction. One species of shark that is currently on the IUCN "Vulnerable" list is the great white shark (*Carcharodon carcharias*). The great white shark is found in coastal surface waters of all the major oceans. It can grow up to 6 m (20 ft.) in length and weigh nearly 2,268 kg (5,000 lb). The great white shark reaches sexual maturity at around 15 years of age and can live for over 30 years. Great white sharks are apex predators (see the figure to the right). An apex predator is an animal that, as an adult, has no natural predators in its ecosystem and resides at the top of the food chain. These sharks prey on marine mammals, fish, and seabirds.

A great white shark

Great white shark conservation has become a global priority in recent years. However, our limited understanding of their behavior has hindered the development of effective conservation strategies for this species. For example, little is known about where and when great white sharks mate, where they give birth, and where they spend their time as juveniles. We also know that some great white sharks travel long distances, such as from Baja California to Hawaii or from South Africa to Australia, but we do not know why they make these journeys. There are, however, a number of potential explanations that have been suggested by scientists. For example, great white sharks might travel long distances because they need to do one or more of the following:

- Find and establish a territory (an area that they defend that contains a mating site and sufficient food resources for them and their young) once they reach sexual maturity or after losing a territory to other great white sharks.
- Migrate between a foraging site and a mating site on an annual or seasonal basis.

LAB 12

- Forage for food—slowly traveling over long distances allows the sharks to find, capture, and consume new sources of food along the way without expending a great deal of energy.
- Find a foraging site with other sharks in it and cooperate with them to capture prey and minimize the amount of energy required to capture and consume food.
- Follow their prey as the prey migrates on an annual or seasonal basis.
- Move between several different foraging areas because they quickly deplete their food source in a given area and must move onto new foraging areas to survive.

All of these potential explanations are plausible because they can help a great white shark survive longer or reproduce more. It is difficult, however, to determine which of these potential explanations is the most valid or acceptable because we know so little about the life history and long-range movements of the great white shark. Most research on this species has been carried out at specific aggregation sites (such as the one near Dyer Island in South Africa). Although this type of research has enabled scientists to learn a lot about the feeding behaviors and short-range movements of the great white shark, we know very little about how they act in other places. A group called OCEARCH (*www.ocearch.org*), however, is trying to facilitate more research on their life history and long-range movements so people can develop better conservation strategies to help protect the great white shark.

This group of researchers has been catching and tagging great white sharks to document where they go over time. To tag and track a great white shark, OCEARCH places a SPOT tag on the shark's dorsal fin. These tags emit a signal that is picked up by global positioning satellites. Unfortunately, the signal can only be detected when the shark's dorsal fin breaks the surface of the water and a satellite is directly overhead. Researchers at OCEARCH call these signals "pings." The time span between pings can vary a great deal (from once an hour to once in a three-week period) because of individual shark behavior and the orbit of a satellite.

OCEARCH has created the Global Shark Tracker database (*www.ocearch.org*) and a companion app for mobile devices (visit the Apple App Store or Google Play to download the free app) to share the real-time data they collect (see the figure on the opposite page). This database allows users to see the current location of all the sharks that the OCEARCH researchers have tagged. It also allows users to track the movement of each shark over time. Users can also search for sharks by name, sex (male or female), and stage of life (mature or immature).

Your Task

Use the OCEARCH Global Shark Tracker database to identify patterns in the long-range movements of the great white shark, and then develop an explanation for those patterns.

Explanations for Animal Behavior
Why Do Great White Sharks Travel Over Long Distances?

A screen shot of Global Shark Tracker from the OCEARCH website

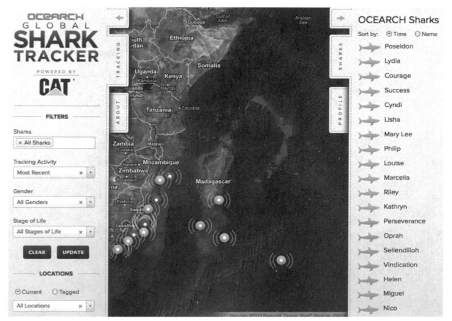

The guiding question of this investigation is, **Why do great white sharks travel over long distances?**

Materials

You will use an online database called Global Shark Tracker to conduct your investigation. You can access the database by going to the following website: *www.ocearch.org*.

Safety Precautions

1. Use caution when working with electrical equipment. Keep away from water sources in that they can cause shorts, fires, and shock hazards. Use only GFI-protected circuits.
2. Wash hands with soap and water after completing this lab.
3. Follow all normal lab safety rules.

Getting Started

Your first step in this investigation is to learn more about what is already known about the great white shark. To do this, check the following websites:

LAB 12

- Animal Diversity Web (*http://animaldiversity.ummz.umich.edu/accounts/Carcharodon_carcharias*)
- MarineBio (*http://marinebio.org/species.asp?id=38*)
- The Smithsonian National Museum of Natural History Ocean Portal (*http://ocean.si.edu/great-white-shark*)

You can then use the OCEARCH Global Shark Tracker database to identify patterns in the long-range movement of great white sharks. To accomplish this task, it is important for you to determine what type of data you will need to collect and how you will analyze it.

To determine *what type of data you will need to collect,* think about the following questions:

- What data will you need to determine if there are patterns in the long-range movements of great white sharks?
- What data will you need to determine if there are sex-related, age-related, or geographic region–related differences in the long-range movements of great white sharks?

To determine *how you will analyze your data,* think about the following questions:

- How can you identify a pattern in the ways great white sharks move over long distances?
- How can you determine if there are patterns in the way great white sharks move over long distances based on sex, age, or geographic region?
- What type of table or graph could you create to help make sense of your data?

Once you have identified patterns in the ways great white sharks move over long distances, you will then need to develop an explanation for those patterns. You can develop one of your own or see if one of the explanations outlined in the "Introduction" section of this investigation is consistent with the patterns you identified. These explanations stem from what scientists know about the behavior of other animals and reflect some of the theories that scientists currently use to explain animal behavior.

Investigation Proposal Required? ☐ Yes ☐ No

Connections to Crosscutting Concepts and to the Nature of Science and the Nature of Scientific Inquiry

As you work through your investigation, be sure to think about

- the importance of identifying patterns,
- the importance of identifying the underlying cause for observations,

- the importance of examining proportional relationships,
- how scientific knowledge can change over time, and
- how the methods used by scientists depend on what is being studied and the research question.

Argumentation Session

Once your group has finished collecting and analyzing your data, prepare a whiteboard that you can use to share your initial argument. Your whiteboard should include all the information shown in the figure to the right.

Argument presentation on a whiteboard

The Guiding Question:	
Our Claim:	
Our Evidence:	Our Justification of the Evidence:

To share your argument with others, we will be using a round-robin format. This means that one member of your group will stay at your lab station to share your group's argument while the other members of your group go to the other lab stations one at a time to listen to and critique the arguments developed by your classmates.

The goal of the argumentation session is not to convince others that your argument is the best one; rather, the goal is to identify errors or instances of faulty reasoning in the arguments so these mistakes can be fixed. You will therefore need to evaluate the content of the claim, the quality of the evidence used to support the claim, and the strength of the justification of the evidence included in each argument that you see. In order to critique an argument, you will need more information than what is included on the whiteboard. You might, therefore, need to ask the presenter one or more follow-up questions, such as:

- Why did you decide to focus on those data?
- What did you do to analyze your data? Why did you decide to do it that way? Did you check your calculations?
- Is that the only way to interpret the results of your analysis? How do you know that your interpretation of your analysis is appropriate?
- Why did your group decide to present your evidence in that manner?
- What other claims did your group discuss before you decided on that one? Why did your group abandon those alternative ideas?
- How confident are you that your claim is valid? What could you do to increase your confidence?

LAB 12

Once the argumentation session is complete, you will have a chance to meet with your group and revise your original argument. Your group might need to gather more data or design a way to test one or more alternative claims as part of this process. Remember, your goal at this stage of the investigation is to develop the most valid or acceptable answer to the research question!

Report

Once you have completed your research, you will need to prepare an investigation report that consists of three sections that provide answers to the following questions:

1. What question were you trying to answer and why?
2. What did you do during your investigation and why did you conduct your investigation in this way?
3. What is your argument?

Your report should answer these questions in two pages or less. This report must be typed, and any diagrams, figures, or tables should be embedded into the document. Be sure to write in a persuasive style; you are trying to convince others that your claim is acceptable or valid!

Lab 12. Explanations for Animal Behavior: Why Do Great White Sharks Travel Over Long Distances?

Checkout Questions

Nowhere in the world is there a movement of animals as spectacular as the wildebeest migration that occurs from July to October each year in Africa. Over 2 million wildebeest travel from Serengeti National Park in Tanzania to the greener pastures of Maasai Mara National Reserve in Kenya. The wildebeest expend a lot of energy to migrate because of the great distance they travel. The wildebeest also have to cross the Mara River in Maasai Mara, where crocodiles will prey on them. In addition, the wildebeest will be hunted, stalked, and run down by the large carnivores found in the Maasai Mara. Many wildebeest, as a result, do not survive the migration.

1. Given the fact that wildebeest must expend a lot of energy and may even die during a migration, why would wildebeest engage in this type of behavior?

2. There is a single, universal, step-by-step scientific method that all scientists follow regardless of the type of question that they are trying to answer.

    a. I agree with this statement.
    b. I disagree with this statement.

    Explain your answer, using examples from your investigation about animal behavior.

LAB 12

3. Scientific knowledge may be abandoned or modified in light of new evidence or because of the reconceptualization of prior evidence and knowledge.

   a. I agree with this statement.
   b. I disagree with this statement.

   Explain your answer, using information from your investigation about animal behavior.

4. Scientists often attempt to identify patterns in nature. Explain why the identification of patterns is useful in science, using an example from your investigation about animal behavior.

5. Scientists often attempt to identify the underlying cause for the observations they make. Explain why the identification of underlying causes is so important in science, using an example from your investigation about animal behavior.

6. Scientists often need to look for proportional relationships. Explain what a proportional relationship is and why these relationships are important, using an example from your investigation about animal behavior.

Lab 13. Environmental Influences on Animal Behavior: How Has Climate Change Affected Bird Migration?

Lab Handout

Introduction

The average temperature in the United States has increased by about 1.3°F since 1910, but the increase in average temperature has not been uniform. Some states have warmed more than others (see the figure below). The *pace* of warming in *all* regions of the United States, however, has accelerated dramatically since the 1970s. This change in pace coincides with the time when the effect of greenhouse gases began to overwhelm the other natural and human influences on climate at the global and continental scales.

A map illustrating how fast each state has been warming each decade since 1970

This increase in average temperature could have a negative impact on many different species of plants and animals because it could lead to changes in seasonal weather patterns, which could then lead to droughts, habitat loss, or food shortages. Migratory birds are one type of animal that may be influenced by a change in climate because birds migrate when

LAB 13

the seasons change. Migratory birds tend to fly north in the spring to breed and return to the warmer wintering grounds of the south when temperatures get colder.

The migration of birds in response to a change of seasons is an example of animal behavior with both a proximate cause and an ultimate cause. A proximate cause is the stimulus that triggers a particular behavior (such as a change in temperature). An ultimate cause, in contrast, is the reason why the behavior exists. In this case, birds migrate because of food and because the longer days of the northern summer provide extended time for breeding birds to feed their young. Migratory birds, as a result, are often able to support larger clutches than nonmigratory species that remain in the tropics year round. This is clearly a benefit of migration.

Environmental conditions serve as both the proximate and ultimate cause of bird migration. Therefore, climate change could have drastic effects on bird migration because it changes seasonal weather patterns. For example, climate change could influence when the temperature drop that serves as the proximate cause of migration for many species of bird happens. Climate change, as noted earlier, can also lead to widespread droughts, habitat loss, and food shortages. These changes in environmental conditions could potentially eliminate the benefits associated with migration because they limit how much access birds have to the resources they need to survive and reproduce after they arrive at their destination.

Your Task

Use the All About Birds website to identify several migratory species of bird that can be found in the United States; then use the eBird online database to determine if the migration behaviors for these species have changed over the last 40 years. If you do see a change, you can then use the National Oceanic and Atmospheric Administration's (NOAA) National Weather Service and National Climatic Data Center databases to explore weather conditions and changes in climate over the same time period.

The guiding question of this investigation is, **How has climate change affected bird migration?**

Materials

You may use any of the following websites during your investigation:

- All About Birds (Cornell Lab of Ornithology): *www.allaboutbirds.org*
- eBird: *http://ebird.org*
- NOAA National Weather Service: *www.weather.gov*
- NOAA National Climatic Data Center: *www.ncdc.noaa.gov*

Environmental Influences on Animal Behavior
How Has Climate Change Affected Bird Migration?

Safety Precautions

1. Use caution when working with electrical equipment. Keep away from water sources in that they can cause shorts, fires, and shock hazards. Use only GFI-protected circuits.
2. Wash hands with soap and water after completing this lab.
3. Follow all normal lab safety rules.

Getting Started

To answer the guiding question, you will need to design and conduct an investigation using three different online databases. Your first step in your investigation, however, is to learn more about birds, why birds migrate, the different migration patterns, and which types of birds migrate. To do this you can visit the website All About Birds, which is sponsored by the Cornell Lab of Ornithology. Your next step is to learn how to use the eBird database to find information on where and when different species of bird have been observed across the United States and over time. You will also need to learn how to use the NOAA National Weather Service database to access information about current weather conditions and the NOAA National Climatic Data Center database to access historical weather conditions for different regions of the United States.

Once you have learned how to use these databases, you will need to determine what type of data you will need to collect, how you will collect it, and how you will analyze it. To determine *what type of data you will need to collect*, think about the following questions:

- How will you determine if there has been a change in bird migration over time?
- What will serve as your dependent variable (e.g., location of breeding and winter locations, abundance of birds, arrival and departure dates in a specific area, distance traveled)?
- What information will you need to be able to link a change in a migration pattern to a change in climate?
- What type of comparisons will you need to make (e.g., different species of bird, birds in different regions, current observations vs. past observations)?

To determine *how you will collect your data*, think about the following questions:

- Where in the eBird and NOAA databases will you look to gather the information you need?
- What tools in the eBird and NOAA databases will you need to use?
- How will you keep track of the data you collect from the three different databases, and how will you organize the data?

LAB 13

To determine *how you will analyze your data,* think about the following questions:

- How will you demonstrate that a change in climate is or is not related to a change in the migration behaviors of bird species?
- How will you quantify a difference or amount of change?
- What type of calculations will you need to make?
- What type of graph could you create to help make sense of your data or to share the data with others?

Investigation Proposal Required?   ☐ Yes   ☐ No

Connections to Crosscutting Concepts and to the Nature of Science and the Nature of Scientific Inquiry

As you work through your investigation, be sure to think about

- the importance of identifying patterns,
- the importance of identifying the underlying cause for observations,
- how systems go through periods of stability and change,
- the nature of scientific knowledge, and
- the difference between data and evidence in science.

Argumentation Session

Once your group has finished collecting and analyzing your data, prepare a whiteboard that you can use to share your initial argument. Your whiteboard should include all the information shown in the figure below.

To share your argument with others, we will be using a round-robin format. This means that one member of your group will stay at your lab station to share your group's argument while the other members of your group go to the other lab stations one at a time to listen to and critique the arguments developed by your classmates.

The goal of the argumentation session is not to convince others that your argument is the best one; rather, the goal is to identify errors or instances of faulty reasoning in the arguments so these mistakes can be fixed. You will therefore need to evaluate the content of the claim, the quality of the evidence used to support the claim, and the strength of the justification of the evidence included in each argument that you see. In order to critique an argument, you will

Argument presentation on a whiteboard

The Guiding Question:	
Our Claim:	
Our Evidence:	Our Justification of the Evidence:

Environmental Influences on Animal Behavior
How Has Climate Change Affected Bird Migration?

need more information than what is included on the whiteboard. You might, therefore, need to ask the presenter one or more follow-up questions, such as:

- Why did you decide to focus on those data?
- What did you do to analyze your data? Why did you decide to do it that way? Did you check your calculations?
- Is that the only way to interpret the results of your analysis? How do you know that your interpretation of your analysis is appropriate?
- Why did your group decide to present your evidence in that manner?
- What other claims did your group discuss before you decided on that one? Why did your group abandon those alternative ideas?
- How confident are you that your claim is valid? What could you do to increase your confidence?

Once the argumentation session is complete, you will have a chance to meet with your group and revise your original argument. Your group might need to gather more data or design a way to test one or more alternative claims as part of this process. Remember, your goal at this stage of the investigation is to develop the most valid or acceptable answer to the research question!

Report

Once you have completed your research, you will need to prepare an investigation report that consists of three sections that provide answers to the following questions:

1. What question were you trying to answer and why?
2. What did you do during your investigation and why did you conduct your investigation in this way?
3. What is your argument?

Your report should answer these questions in two pages or less. This report must be typed, and any diagrams, figures, or tables should be embedded into the document. Be sure to write in a persuasive style; you are trying to convince others that your claim is acceptable or valid!

Lab 13. Environmental Influences on Animal Behavior: How Has Climate Change Affected Bird Migration?

Checkout Questions

Use the following information to answer questions 1–3. A biologist is interested in mealworm behavior. He sets up a box as shown below. He uses a neon lamp for light and constantly waters pieces of paper for moisture. In the center of the box he places 20 mealworms.

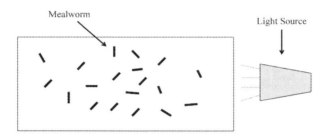

One day later he returns to see where the mealworms ended up.

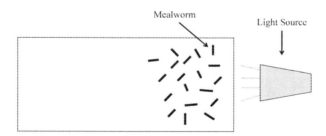

Biologists use the concepts of proximate and ultimate cause for behavior to explain this type of observation.

1. Describe the concept of a proximate cause of a behavior.

Environmental Influences on Animal Behavior
How Has Climate Change Affected Bird Migration?

2. Describe the concept of an ultimate cause of a behavior.

3. Use the concepts of proximate and ultimate causes of a behavior to explain why the mealworms moved to one side of the box.

4. All scientific knowledge is certain and does not change.

    a. I agree with this statement.
    b. I disagree with this statement.

    Explain your answer, using examples from your investigation about animal behavior.

5. Evidence is data that are used to support a claim.

    a. I agree with this statement.
    b. I disagree with this statement.

    Explain your answer, using information from your investigation about animal behavior.

LAB 13

6. Scientists often attempt to identify patterns in nature. Explain why the identification of patterns is useful in science, using an example from your investigation about animal behavior.

7. Scientists often attempt to identify the underlying cause for the observations they make. Explain why the identification of underlying causes is so important in science, using an example from your investigation about animal behavior.

8. Biological systems often go through periods of stability and change. Explain what this means, using an example from your investigation about animal behavior.

Lab 14. Interdependence of Organisms: Why Is the Sport Fish Population of Lake Grace Decreasing in Size?

Lab Handout

Introduction

Lake Grace (see the figure to the right) is known as one of the best lakes for sport fishing in the United States. The Tolt and Faith rivers feed the lake, and extensive stump and grass beds provide a great habitat for sport fish, such as largemouth bass, white bass, and bluegill. Sizable populations of other fish, such as catfish, crappie, and bream, are also present. In fact, over 79 different species of fish have been found in the lake. Over the last five years, however, anglers have been catching fewer and fewer of the large sport fish that once made Lake Grace so famous.

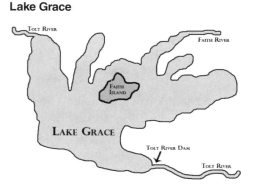

Lake Grace

The low numbers of sport fish in the lake have led to a decrease in the number of anglers that come to the lake to fish on weekends or for a fishing vacation. As a result, there has been a downturn in the economy of the nearby town of Aidanville, and many local stores and hotels that depended on tourism have gone out of business.

Your Task

Conduct an investigation of the water quality of Lake Grace and develop an explanation for the decline in the populations of sport fish.

The guiding question of this investigation is, **Why is the sport fish population of Lake Grace decreasing in size?**

Materials

You may use any of the following materials during your investigation:

- Samples of water from Lake Grace (three different locations)
- Water quality test kit (pH, nitrates, phosphates, dissolved oxygen, turbidity)
- Information packet

LAB 14

Safety Precautions

1. Safety goggles, vinyl gloves, and aprons are required for this activity.
2. Wash hands with soap and water upon completing this lab.
3. Follow all normal lab safety rules.

Getting Started

To answer the guiding question, you will need to analyze an existing data set and then determine the overall quality of a water sample from Lake Grace. To accomplish this task, you must first determine what type of data you will need to collect, how you will collect it, and how you will analyze it. To determine *what type of data you will need to collect,* think about the following questions:

- What type of information do I need to collect from the existing data set found in the information packet?
- What type of tests will I need to determine the quality of the water in Lake Grace? (*Hint*: Be sure to follow all directions as given in the water quality test kits.)
- What type of measurements or observations will you need to record during your investigation?

To determine *how you will collect your data,* think about the following questions:

- What will serve as a control (or comparison) condition?
- How will you make sure that your data are of high quality (i.e., how will you reduce error)?
- How will you keep track of the data you collect and how will you organize the data?

To determine *how you will analyze your data,* think about the following questions:

- What type of calculations will you need to make?
- What type of graph could you create to help make sense of your data?

Investigation Proposal Required?   ☐ Yes   ☐ No

Connections to Crosscutting Concepts and to the Nature of Science and the Nature of Scientific Inquiry

As you work through your investigation, be sure to think about

- the importance of identifying the underlying cause for observations;

Interdependence of Organisms
Why Is the Sport Fish Population of Lake Grace Decreasing in Size?

- why it is important to determine what is relevant at a particular scale or over a specific time frame;
- how energy and matter flow into, out of, within, and through a system;
- how the method scientists use depends on the topic under investigation and the research question; and
- how social and cultural issues influence the work of scientists.

Argumentation Session

Once your group has finished collecting and analyzing your data, prepare a whiteboard that you can use to share your initial argument. Your whiteboard should include all the information shown in the figure to the right.

Argument presentation on a whiteboard

The Guiding Question:	
Our Claim:	
Our Evidence:	Our Justification of the Evidence:

To share your argument with others, we will be using a round-robin format. This means that one member of your group will stay at your lab station to share your group's argument while the other members of your group go to the other lab stations one at a time to listen to and critique the arguments developed by your classmates.

The goal of the argumentation session is not to convince others that your argument is the best one; rather, the goal is to identify errors or instances of faulty reasoning in the arguments so these mistakes can be fixed. You will therefore need to evaluate the content of the claim, the quality of the evidence used to support the claim, and the strength of the justification of the evidence included in each argument that you see. In order to critique an argument, you will need more information than what is included on the whiteboard. You might, therefore, need to ask the presenter one or more follow-up questions, such as:

- How did you collect your data? Why did you use that method? Why did you collect those data?
- What did you do to make sure the data you collected are reliable? What did you do to decrease measurement error?
- What did you do to analyze your data? Why did you decide to do it that way? Did you check your calculations?
- Is that the only way to interpret the results of your analysis? How do you know that your interpretation of your analysis is appropriate?
- Why did your group decide to present your evidence in that manner?

LAB 14

- What other claims did your group discuss before you decided on that one? Why did your group abandon those alternative ideas?
- How confident are you that your claim is valid? What could you do to increase your confidence?

Once the argumentation session is complete, you will have a chance to meet with your group and revise your original argument. Your group might need to gather more data or design a way to test one or more alternative claims as part of this process. Remember, your goal at this stage of the investigation is to develop the most valid or acceptable answer to the research question!

Report

Once you have completed your research, you will need to prepare an investigation report that consists of three sections that provide answers to the following questions:

1. What question were you trying to answer and why?
2. What did you do during your investigation and why did you conduct your investigation in this way?
3. What is your argument?

Your report should answer these questions in two pages or less. This report must be typed, and any diagrams, figures, or tables should be embedded into the document. Be sure to write in a persuasive style; you are trying to convince others that your claim is acceptable or valid!

Interdependence of Organisms
Why Is the Sport Fish Population of Lake Grace Decreasing in Size?

Lake Grace Information Packet

Lake Grace and the Town of Aidanville

Lake Grace is located in the southeastern United States and covers an area of 37,500 acres. Extending up the Tolt River 30 miles and up the Faith River 35 miles, Lake Grace has 376 miles of shoreline. The lake was created in 1957 when the Tolt River Dam was built. The dam produces hydroelectric power that is used by both homes and industry in the area. Aidanville was founded in 1897 and is located on the southwest side of Lake Grace (see the figure below).

Lake Grace and environs

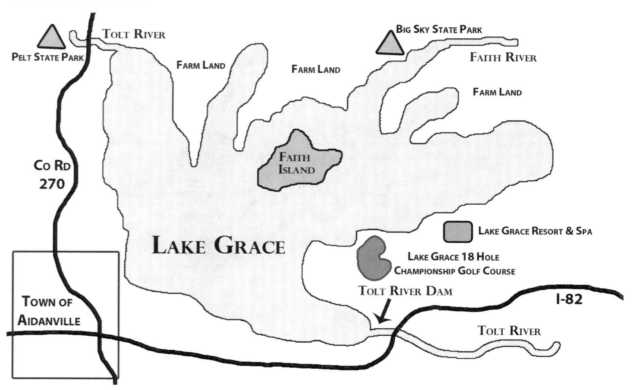

LAB 14

Major events in the history of Lake Grace and the town of Aidanville

Date	Event
1940	The population of Aidanville reaches 1,587, according to U.S. census data.
1947	Money to build the Tolt River Dam is authorized, and construction begins.
1952	The Tolt River Dam is completed.
1957	Lake Grace opens for public use.
1958	Pelt State Park and Big Sky State Park are completed and open for public use.
1980	The population of Aidanville reaches 2,016, according to U.S. census data.
1981	Two Rivers State Park is completed and opens for use. The park includes a public boat launch, which was built in response to the large number of anglers coming to the lake to fish.
1985	Three new hotels are built in Aidanville.
1989	Two more hotels and five more restaurants are built in Aidanville.
1990	The population of Aidanville reaches 3,287, according to U.S. census data.
1995	Aidanville City Council begins a program to monitor the water quality of Lake Grace.
1996	Lake Grace Resort and Spa completed
1998	Invasive species of water plants, such as hydrilla and water hyacinth, are found in Lake Grace for the first time.
1999	Lake Grace 18-hole championship golf course is completed and open for public use.
2000	The population of Aidanville reaches 3,824, according to U.S. census data.
2001	The farmers who own the farmland near Big Sky State Park stop raising crops, sell off part of their land to developers, and begin operating a large hog farm.
2004	The Aidanville City Council begins to use herbicides to slow the spread of invasive water plants in Lake Grace.
2011	Three hotels and four restaurants go out of business in Aidanville.

Interdependence of Organisms
Why Is the Sport Fish Population of Lake Grace Decreasing in Size?

Number and size of sport fish caught annually in Lake Grace, 1995–2011

Year	Number caught			Average size of fish caught (cm)		
	Largemouth bass	White bass	Bluegill	Largemouth bass*	White bass†	Bluegill‡
2011	1,152	1,705	952	31	39	17
2010	1,287	1,830	975	29	38	19
2009	1,213	1,819	1,012	30	38	16
2008	1,284	1,962	1,204	32	39	19
2007	1,406	1,993	1,432	31	42	20
2006	1,517	2,003	1,616	30	43	21
2005	1,872	1,894	2,203	33	41	22
2004	2,411	1,752	2,106	32	45	24
2003	1,310	1,385	1,910	33	49	25
2002	1,504	1,206	1,867	34	48	23
2001	1,692	1,197	1,992	33	43	28
2000	1,825	1,151	1,845	36	45	27
1999	1,714	1,302	1,791	38	47	31
1998	1,535	1,207	1,603	39	49	29
1997	2,387	1,234	1,375	40	43	27
1996	1,747	1,750	1,402	43	48	32
1995	2,422	1,344	1,208	41	46	33

* The legal size limit for largemouth bass is 30 cm.

† The legal size limit for white bass is 40 cm.

‡ There is no legal size limit for bluegill.

LAB 14

Information about some of the organisms found in Lake Grace

Name	Appearance	Habitat	Size	Diet	Reproduction	Ecological importance
Largemouth bass		Shallow lakes, ponds, or rivers	Adults range in size from 26 cm to 46 cm	Young fish feed on daphnia, gammarus amphipods, and invertebrates; adults feed on small fish, frogs, and aquatic invertebrates	Male fish build nests in sand or gravel in shallow areas and attract a female to the nest. The female lays a few hundred eggs and then the male fertilizes them. The male guards the eggs until they hatch, which takes 7–10 days.	Largemouth bass are important predators in freshwater ecosystems and help maintain the population size of other primary consumers such as aquatic invertebrates and secondary consumers (such as amphibians and small fish). They are also one of the species most sought after by recreational anglers.
White bass		Deep, clear lakes and large rivers.	Adults range in size from 38 cm to 50 cm	Mostly daphnia, crustaceans, and other aquatic invertebrates; larger individuals feed on small fish	They only spawn in water ranging from 12 to 20°C. Females lay eggs in moving water such as a tributary stream or river. Females release 200,000 eggs, which stick to the surface of plants, submerged logs, gravel, or rocks.	White bass are important predators in freshwater ecosystems and help maintain the population size of other primary consumers such as aquatic invertebrates and secondary consumers (such as amphibians and small fish). They are also one of the species most sought after by recreational anglers.
Bluegill		Weedy, shallow, waters; does not tolerate high turbidity well	Adults range in size from 15 cm to 35 cm	Daphnia, gammarus amphipods, insects, and crustaceans	They spawn early in the spring. Females release their eggs and males then fertilize them. Eggs hatch about eight days later.	Bluegills are important aquatic predators. They also provide food for larger fish. Numerous organisms eat their eggs. For anglers, the bluegill provides considerable sport, and the flesh is firm, flaky and well flavored. Bluegills are often stocked in artificial ponds as forage for largemouth bass.
Daphnia		Lakes with temperatures below 20°C	1 mm long	Bacteria, protists, and algae	They produce eggs that develop without fertilization. An adult female can produce 10–15 eggs.	Daphnia are a principal food staple for fish and an important link in the food chain (fish stomach can contain 95% daphnia by volume). Daphnia also help maintain water quality by cleaning up algae blooms in lakes (daphnia can reduce the amount of algae in a lake by half in a small amount of time).

Interdependence of Organisms
Why Is the Sport Fish Population of Lake Grace Decreasing in Size?

Information about some of the organisms found in Lake Grace *(continued)*

Name	Appearance	Habitat	Size	Diet	Reproduction	Ecological importance
Gammarus amphipod		Floors of lakes and rivers that are well oxygenated and below 20°C	21 mm long	Algae and dead organic matter	Occurs during winter—females only produce one brood during their life (which lasts 1–1.5 years)	Gammarus amphipods are a main source of food for larger freshwater organisms. Gammarus amphipods are sensitive to changes in the environment—low pH levels or warm temperatures kill them.
Algae		Anywhere there is a body of water or a sufficient quantity of moisture	Can live as single cells, in colonies (groups), or as strands of attached cells (called filaments)	Photosynthetic organism	Asexual	When the concentrations of nitrates and phosphates increase in a lake, the algae population increases. The massive amount of algae gives the water a pea-green color and produces a funny smell (called an algae bloom). Many of the algae begin to die off as the population increases. Oxygen-using decomposer bacteria then increase in number, which drops the oxygen levels of the lake. As a result, many fish can suffocate.
Pickerelweed		Lakes, ponds, ditches, and streams	60–90 cm tall	Photosynthetic organism	Seeds	Submerged portions of aquatic plants provide habitats for many invertebrates. These invertebrates in turn are used as food by fish and other wildlife species (e.g., amphibians, reptiles, ducks). After aquatic plants die, their decomposition by bacteria and fungi provides food for many aquatic invertebrates.
Hydrilla		Lakes, ponds, ditches, and streams	Up to 760 cm tall	Photosynthetic organism	Seeds and fragmentation	Grows into dense stands extending from the shoreline to a depth of 10 ft. Dense strands can (1) prevent light from penetrating to deeper water, (2) reduce dissolved oxygen, and (3) displace native plants and reduce biodiversity.
Water hyacinth		Lakes, ponds, ditches, and streams; floats above the water surface	Leaves are 10–20 cm and can be up to 1 m above the surface	Photosynthetic organism	Seeds	Water hyacinth will cover lakes and ponds entirely; this has a dramatic impact on water flow, blocks sunlight from reaching native aquatic plants, and starves the water of oxygen. The plants also create a prime habitat for mosquitoes.

LAB 14

Water quality in Lake Grace

The following table provides data from a program that was started in 1995 by the Aidanville City Council to monitor the water quality in Lake Grace. Unfortunately, the program was cut in 2005 because the city lacked the funds necessary to sustain it.

Year	pH	Dissolved oxygen (mg/L)	Nitrates* (ppm)	Phosphate† (mg/L)	Coliform bacteria	Triclopyr‡ (ppb)	Algal bloom§ observed
2005	5.8	9.5	36	0.12	Yes	11	No
2004	5.9	9.5	35	0.13	Yes	8	No
2003	5.9	9.6	33	0.12	Yes	1	No
2002	6.0	9.6	30	0.11	Yes	2	Yes
2001	6.0	9.7	12	0.12	No	2	No
2000	6.0	9.6	13	0.1	No	2	Yes
1999	6.2	9.8	12	0.09	No	2	No
1998	6.4	10.3	14	0.02	No	1	No
1997	6.6	10.2	13	0.01	No	1	No
1996	6.5	10.2	14	0.02	No	2	No
1995	6.7	10.2	12	0.01	No	1	No

* Nitrate levels over 30 ppm can inhibit growth of fish and some aquatic invertebrates and stimulate the growth of algae and other aquatic plants

† Phosphate levels of 0.01–0.03 mg/L in lake water are considered normal. Plant growth is stimulated at levels of 0.025–0.1 mg/L.

‡ Triclopyr is a weed killer (herbicide) that targets broadleaf plants and has often been used in in lakes. Levels of triclopyr, however, must be below 2 ppb for the water to be safe for irrigation; higher doses of the chemical can be toxic to aquatic organisms. Small organisms such as gammarus amphipods, daphnia, and freshwater shrimp often ingest tiny amounts of the chemical into their bodies or absorb small amounts of it through their gills, but the small dose of the chemical usually does little harm to these organisms (although the chemical often stays in their system for long periods of time).

§ An algal bloom is a rapid increase or accumulation in the population of algae. Although there is no officially recognized threshold level, algae can be considered to be blooming at concentrations of hundreds to thousands of cells per milliliter of water. Algal bloom concentrations may reach millions of cells per milliliter of water. Algal blooms are the result of an excess of nutrients, particularly phosphorus and nitrates. The excess of nutrients may originate from fertilizers that are applied to land for agricultural or recreational purposes (such as keeping the grass on fairways healthy). These nutrients can then enter rivers and lakes through water runoff. When phosphates and nitrates are introduced into water systems, higher concentrations cause increased growth of algae and plants. Algae tend to grow very quickly under high nutrient availability, but each alga is short-lived, and the result is a high concentration of dead organic matter, which starts to decay. The decay process consumes dissolved oxygen in the water.

Interdependence of Organisms
Why Is the Sport Fish Population of Lake Grace Decreasing in Size?

Average water temperature in Lake Grace

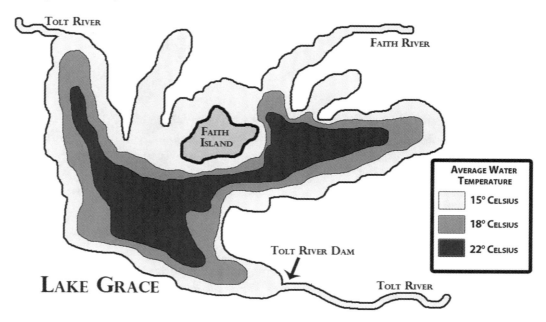

LAB 14

Lab 14. Interdependence of Organisms: Why Is the Sport Fish Population of Lake Grace Decreasing in Size?

Checkout Questions

1. Living organisms in an ecosystem interact with each other. If one of those organisms is removed from the ecosystem, how does that affect the remaining organisms? Use a specific example from the Lake Grace scenario as part of your explanation.

2. Scientists do not study topics or problems that are important to everyday life because science is independent of society and culture.

    a. I agree with this statement.
    b. I disagree with this statement.

    Explain your answer, using examples from your investigation about Lake Grace.

3. Scientists rely on many different types of methods such as experiments, fieldwork, systematic observations, and the analysis of an existing data set.

    a. I agree with this statement.
    b. I disagree with this statement.

    Explain your answer, using examples from your investigation about Lake Grace.

4. Scientists often attempt to identify the underlying cause for the observations they make. Explain why the identification of underlying causes is so important in science, using an example from your investigation about Lake Grace.

5. Scientists often attempt to track how energy and matter flow into, out of, and within a system during an investigation. Explain why tracking the flow of energy and matter is useful in science, using an example from your investigation about Lake Grace.

6. Scientists often need to determine what is and what is not relevant at a particular scale or during a particular time frame during an investigation. Explain why this is so important, using an example from your investigation about Lake Grace.

LAB 15

Lab 15. Competition for Resources: How Has the Spread of the Eurasian Collared-Dove Affected Different Populations of Native Bird Species?

Lab Handout

Introduction

A community is any assemblage of populations in an area or a habitat. There are a number of different interspecies interactions that take place within a community. One example of an interaction that takes place between species is competition. Organisms compete for resources, such as food, water, and space, when resources are in short supply. For example, weeds and grass compete for soil nutrients and water, grasshoppers and bison compete for grass, and lynx and foxes compete for hares. There is potential for competition between any two species populations that need the same limited resource. Resources, however, are not always scarce in every community (e.g., water in the ocean or oxygen on the Great Plains). Species therefore do not always compete for every resource they need to survive.

Species also do not compete for resources when they occupy different ecological niches. An ecological niche is the sum total of a species' use of biotic and abiotic resources in its environment. An organism's ecological niche is its ecological role or how it fits into an ecosystem. The ecological niche of a bird, for example, includes the temperature range it tolerates, the type(s) of tree it nests in, the material it uses to build its nest, the time of day it is active, and the type of insects or seeds it eats (along with numerous other components). Species with different ecological niches require different resources and play different roles in a community. Therefore, species with different ecological niches rarely compete for the same resources.

In the first half of the 20th century, a Russian ecologist named Georgii Gause formulated a law known as the competitive exclusion principle. This law states that two species occupying the same ecological niche cannot coexist in the same community because one will use the resources more efficiently and thus be able to reproduce more offspring. The reproductive advantage of one species results in the local elimination of the other one. Competitive exclusion, as a result, can have a significant effect on the number and types of organisms found within an ecosystem. It is important to note, however, that species with similar (but not identical) ecological niches can coexist in the same community if there is at least one significant difference in their niches (such as when they are active or what they eat).

Invasive species are organisms that are not native to an ecosystem. These organisms are introduced into a new environment through some type of human activity. Invasive species often colonize a community and spread rapidly. They are able to colonize and spread

Competition for Resources
How Has the Spread of the Eurasian Collared-Dove Affected Different Populations of Native Bird Species?

because they can tolerate a wide variety of habitat conditions; they grow fast, reproduce often, compete aggressively for resources, and usually lack natural enemies in the new community. Invasive species, as a result, can cause environmental, economic, and human harm by displacing native species, altering habitats, upsetting the balance of an ecosystem, or degrading the quality of recreation areas.

An example of an invasive species is the Eurasian collared-dove (see the figure to the left). This bird was introduced to the Bahamas in 1970 and spread from there to Florida in 1982. It has since spread across North America and is now found as far south as Veracruz, as far west as California, and as far north as Alaska (see the figure on the next page). Although the Eurasian collared-dove does not migrate, it spreads and then colonizes new areas at an alarming rate. In Arkansas, for example, it took only five years (1997–2002) for it to spread from the southeast corner of the state to the northwest corner (a distance of about 500 km).

The Eurasian collared-dove

The impact of the Eurasian collared-dove on native bird species in North America is not yet known, but it seems to occupy an ecological niche that is similar to the other members of the dove family (Columbidae). Scientists are attempting to determine if the Eurasian collared-dove will outcompete native dove species for available resources. They are also interested in the impact that this invasive species may have on other native species of nonmigratory bird. Fortunately, there are a number of databases that allow scientists to track where different species of bird can be found, when they can be found, and how common they are in a given location. One such database is eBird, which enables users to go online to access observational data submitted by bird-watchers at thousands of locations across the United States. Scientists can use these data and the visualization tools built into the website to examine the frequency and abundance of different species of birds at different locations and over time.

Your Task

Identify at least two native bird species that occupy a similar ecological niche as the Eurasian collared-dove. Then determine what has happened to these two species of birds over time as the Eurasian collared-dove spread across the United States.

The guiding question of this investigation is, **How has the spread of the Eurasian collared-dove affected different populations of native bird species?**

Materials

You may use any of the following resources during your investigation:

- eBird database: *http://ebird.org*
- All About Birds database: *www.allaboutbirds.org*

LAB 15

Range maps of the Eurasian collared-dove in the United States (shaded areas indicate a sighting during that time period)

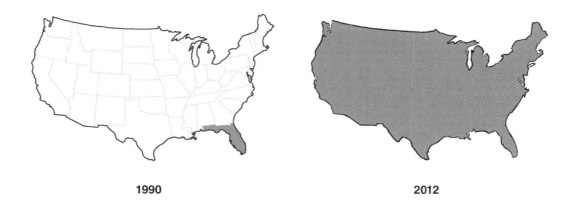

1990 2012

Safety Precautions

1. Use caution when working with electrical equipment. Keep away from water sources in that they can cause shorts, fires, and shock hazards. Use only GFI-protected circuits.

2. Wash hands with soap and water after completing this lab.

3. Follow all normal lab safety rules.

Getting Started

To answer the guiding question, you will need to design and conduct an investigation. To accomplish this task, you must determine what type of data you will need to collect, how you will collect it, and how you will analyze it. To determine *what type of data you will need to collect*, think about the following questions:

- Which native bird species will you include in your investigation?
- How will you determine if a native species of bird has been affected by the spread of the Eurasian collared-dove?
- What will serve as your dependent variable (e.g., range, frequency of sightings, abundance)?
- What type of observations will you need to record during your investigation?

To determine *how you will collect your data*, think about the following questions:

Competition for Resources
How Has the Spread of the Eurasian Collared-Dove Affected Different Populations of Native Bird Species?

- What will you need to compare?
- How often will you collect data and when will you do it?
- How will you keep track of the data you collect and how will you organize the data?

To determine *how you will analyze your data,* think about the following questions:

- What type of calculations will you need to make?
- What type of graph could you create to help make sense of your data?

Investigation Proposal Required?   ☐ Yes   ☐ No

Connections to Crosscutting Concepts and to the Nature of Science and the Nature of Scientific Inquiry

As you work through your investigation, be sure to think about

- the importance of identifying and explaining patterns,
- what is relevant or important at different times and scales,
- the difference between observations and inferences in science, and
- how society and cultural values shape the work of scientists.

Argumentation Session

Once your group has finished collecting and analyzing your data, prepare a whiteboard that you can use to share your initial argument. Your whiteboard should include all the information shown in the figure below.

Argument presentation on a whiteboard

The Guiding Question:	
Our Claim:	
Our Evidence:	Our Justification of the Evidence:

To share your argument with others, we will be using a round-robin format. This means that one member of your group will stay at your lab station to share your group's argument while the other members of your group go to the other lab stations one at a time to listen to and critique the arguments developed by your classmates.

The goal of the argumentation session is not to convince others that your argument is the best one; rather, the goal is to identify errors or instances of faulty reasoning in the arguments so these mistakes can be fixed. You will therefore need to evaluate the content of the claim, the quality of the evidence used to support the claim, and the strength of the justification of the evidence included in each argument that you see. In order to critique an argument, you will need more information than what is included on the whiteboard. You might, therefore, need to ask the presenter one or more follow-up questions, such as:

- Why did you decide to focus on those data?
- What did you do to analyze your data? Why did you decide to do it that way? Did you check your calculations?
- Is that the only way to interpret the results of your analysis? How do you know that your interpretation of your analysis is appropriate?
- Why did your group decide to present your evidence in that manner?
- What other claims did your group discuss before you decided on that one? Why did your group abandon those alternative ideas?
- How confident are you that your claim is valid? What could you do to increase your confidence?

Once the argumentation session is complete, you will have a chance to meet with your group and revise your original argument. Your group might need to gather more data or design a way to test one or more alternative claims as part of this process. Remember, your goal at this stage of the investigation is to develop the most valid or acceptable answer to the research question!

Report

Once you have completed your research, you will need to prepare an investigation report that consists of three sections that provide answers to the following questions:

1. What question were you trying to answer and why?
2. What did you do during your investigation and why did you conduct your investigation in this way?
3. What is your argument?

Your report should answer these questions in two pages or less. This report must be typed, and any diagrams, figures, or tables should be embedded into the document. Be sure to write in a persuasive style; you are trying to convince others that your claim is acceptable or valid!

Lab 15. Competition for Resources: How Has the Spread of the Eurasian Collared-Dove Affected Different Populations of Native Bird Species?

Checkout Questions

Use the following information to answer questions 1 and 2. A biologist sets up three test tubes with cultures of microorganisms: test tube A contains *Paramecium aurelia,* test tube B contains *Paramecium caudatum,* and test tube C contains both *P. aurelia* and *P. caudatum.* She then records how the size of each population changes over time in each test tube. The data she collected are provided in the figure below.

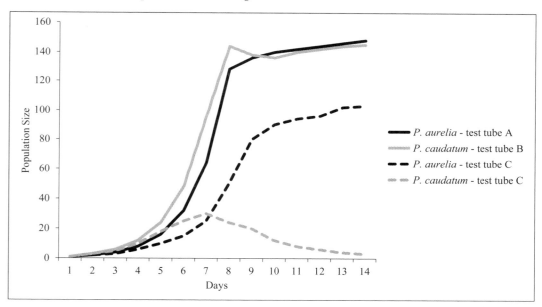

Biologists often use the competitive exclusion principle to explain these types of observations in other contexts.

1. Describe the competitive exclusion principle.

LAB 15

2. Use the competitive exclusion principle to explain these observations.

3. Scientists create laws.

   a. I agree with this statement.
   b. I disagree with this statement.

   Explain your answer, using information from your investigation about competition for resources.

4. Scientific research is not influenced by social and cultural values or expectations because scientists are trained to conduct unbiased studies.

   a. I agree with this statement.
   b. I disagree with this statement.

   Explain your answer, using examples from your investigation about competition for resources.

Competition for Resources
How Has the Spread of the Eurasian Collared-Dove Affected Different Populations of Native Bird Species?

5. Scientists often attempt to identify patterns in nature. Explain why the identification of patterns is useful in science, using an example from your investigation about competition for resources.

6. Scientists often need to determine what is and what is not relevant at different times and scales during an investigation. Explain why this is important to do, using an example from your investigation about competition for resources.

SECTION 4
Life Sciences Core Idea 3:

Heredity: Inheritance and Variation of Traits

Lab 16. Mendelian Genetics: Why Are the Stem and Leaf Color Traits of the Wisconsin Fast Plant Inherited in a Predictable Pattern?

Lab Handout

Introduction

When dogs are bred, the result is puppies, and when racehorses are bred, the result is a foal. The same is true for plants. When one pea plant fertilizes another pea plant, each seed that is produced will become a pea plant and not a tulip, a rose, or a geranium. Species produce more of their own kind because each species passes down a specific set of traits from generation to generation. These traits make each species unique. But to anyone who has bred racehorses, dogs, or pea plants, it is abundantly clear that there are differences among members of the same species. Where do these variations come from, and how are these traits passed on from parents to offspring? These questions baffled scientists for hundreds of years, until Gregor Mendel was able to explain how traits are inherited.

Gregor Mendel identified the rules that govern heredity by crossing (breeding) individual pea plants with different versions of a trait and then documenting which version of that trait was inherited by the offspring. He also tracked the inheritance of specific versions of a trait over many generations. Once he gathered enough data, he was able to develop a set of rules that he could use to predict the traits of an offspring based on the traits of the parents. These rules are now known as Mendel's model of inheritance and are still used by scientists and medical doctors today. In this investigation, your goal will be to develop a model of inheritance that explains how traits are passed on from parent to offspring, much like Mendel did.

Your Task

Use a computer simulation to cross Wisconsin Fast Plants with different traits in order to identify patterns in the ways these traits are inherited. Once you have identified these patterns, you will need to develop a model that explains how traits are inherited in this organism. You will then need to test your model to determine how well it allows you to predict the traits of offspring.

The guiding question of this investigation is, **Why are the stem and leaf color traits of the Wisconsin Fast Plant inherited in a predictable pattern?**

LAB 16

Materials

You will use an online simulation called *Observing One or Two Traits in Wisconsin Fast Plants* to conduct your investigation. You can access the simulation by going to the following web page: *www.fastplants.org/legacy/genetics/Introductions/two-trait.htm*.

Safety Precautions

1. Use caution when working with electrical equipment. Keep away from water sources in that they can cause shorts, fires, and shock hazards. Use only GFI-protected circuits.

2. Wash hands with soap and water after completing this lab.

3. Follow all normal lab safety rules.

Getting Started

Your goal for this investigation is to develop a model that explains why the traits of the Wisconsin Fast Plant are inherited in a predictable pattern. Biologists determine how traits are passed from parent to offspring by (1) crossing (breeding) two individuals with specific traits (e.g., plants with purple or green stems and plants with dark green or light green leaves) and then recording the traits of their offspring in the next generation, (2) looking for patterns in the way specific traits are passed down from generation to generation, and (3) using these data to generate a model that explains why they are inherited in this manner. They then test their model to see how well it can predict the outcome of other crosses.

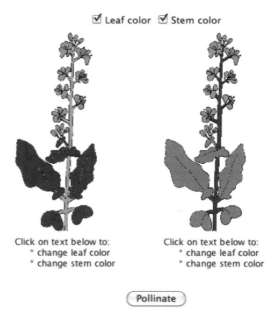

A screen shot from the *Observing One or Two Traits in Wisconsin Fast Plants* simulation

In the field, this type of research can be slow because a single generation can take anywhere from several weeks to several years. Fortunately, we can speed up the process of developing an explanatory model inside the classroom by using a computer simulation that allows you to breed pea plants and observe the results of your crosses in a matter of seconds.

Mendelian Genetics
Why Are the Stem and Leaf Color Traits of the Wisconsin Fast Plant Inherited in a Predictable Pattern?

The *Observing One or Two Traits in Wisconsin Fast Plants* simulation (see the figure on the opposite page) enables you to do the following:

- Choose if you want to look at stem and/or leaf color of pea plants.
- Cross (breed) parents with different traits. (To cross the plants, click on the "Pollinate" button. You will see a bee travel between the two plants to indicate that the plants are being crossed.)
- See the 64 progeny (offspring) of the two plants that you crossed.
- Choose any two plants from the 64 progeny and cross them as well.
- Go back to any screen and repeat a cross. They will be labeled F2a, F2b, F2c, and so on.
- Perform a test cross with a parent plant and one of its offspring. A test cross involves breeding an individual of the dominant phenotype with an recessive phenotype individual to see what genotype the dominant parent is based on the trait ratios of the offspring

Remember, your first step in this investigation is to use this online simulation to identify patterns in the way the steam color and leaf color traits are inherited in Wisconsin Fast Plants. Once you have identified some patterns, you need to develop a model that you can use to explain them. Your model, like Mendel's, will likely consist of several different postulates. A postulate is a tentative claim. Mendel's model of inheritance included four postulates. The first postulate was, "Specific traits are determined by an inheritable unit that is passed down from parent to offspring." You can use the first postulate in Mendel's model of inheritance as one of the postulates in your model of inheritance.

Once you have developed your model, you will need to determine if it is valid or not by testing it with the online simulation. You will know when your model is valid because a valid model allows you to predict the traits of offspring. Your model, in other words, should enable you to predict the traits of the next generation of plants based on the traits of the parent plants. If your model allows you to make accurate predictions, then it is valid. If your model results in inaccurate predictions, then it is flawed and will need to be changed.

Investigation Proposal Required?   ☐ Yes   ☐ No

Connections to Crosscutting Concepts and to the Nature of Science and the Nature of Scientific Inquiry

As you work through your investigation, be sure to think about

- the importance of identifying patterns,
- how scientists attempt to uncover causal mechanisms,
- the value of looking at proportional relationships when analyzing data,

LAB 16

- the difference between theories and laws in science, and
- the importance of imagination and creativity in science.

Argumentation Session

Once your group has finished collecting and analyzing your data, prepare a whiteboard that you can use to share your initial argument. Your whiteboard should include all the information shown in the figure below.

Argument presentation on a whiteboard

The Guiding Question:	
Our Claim:	
Our Evidence:	Our Justification of the Evidence:

To share your argument with others, we will be using a round-robin format. This means that one member of your group will stay at your lab station to share your group's argument while the other members of your group go to the other lab stations one at a time to listen to and critique the arguments developed by your classmates.

The goal of the argumentation session is not to convince others that your argument is the best one; rather, the goal is to identify errors or instances of faulty reasoning in the arguments so these mistakes can be fixed. You will therefore need to evaluate the content of the claim, the quality of the evidence used to support the claim, and the strength of the justification of the evidence included in each argument that you see. In order to critique an argument, you will need more information than what is included on the whiteboard. You might, therefore, need to ask the presenter one or more follow-up questions, such as:

- How did you use the simulation to collect your data?
- What did you do to analyze your data? Why did you decide to do it that way? Did you check your calculations?
- Is that the only way to interpret the results of your analysis? How do you know that your interpretation of your analysis is appropriate?
- Why did your group decide to present your evidence in that manner?
- What other claims did your group discuss before you decided on that one? Why did your group abandon those alternative ideas?
- How confident are you that your claim is valid? What could you do to increase your confidence?

Once the argumentation session is complete, you will have a chance to meet with your group and revise your original argument. Your group might need to gather more data or design a way to test one or more alternative claims as part of this process. Remember, your

goal at this stage of the investigation is to develop the most valid or acceptable answer to the research question!

Report

Once you have completed your research, you will need to prepare an investigation report that consists of three sections that provide answers to the following questions:

1. What question were you trying to answer and why?
2. What did you do during your investigation and why did you conduct your investigation in this way?
3. What is your argument?

Your report should answer these questions in two pages or less. This report must be typed, and any diagrams, figures, or tables should be embedded into the document. Be sure to write in a persuasive style; you are trying to convince others that your claim is acceptable or valid!

LAB 16

Lab 16. Mendelian Genetics: Why Are the Stem and Leaf Color Traits of the Wisconsin Fast Plant Inherited in a Predictable Pattern?

Checkout Questions

1. Describe Mendel's model of inheritance.

2. Use Mendel's model of inheritance to explain why traits can skip a generation.

3. Scientists do not use their imagination and creativity at any time during an investigation because science is based on logic and reason.

   a. I agree with this statement.
   b. I disagree with this statement.

   Explain your answer, using examples from your investigation about Mendelian genetics.

4. A scientific law is a description of a relationship or a pattern in nature (i.e., laws describe *how* things happen).

Mendelian Genetics
Why Are the Stem and Leaf Color Traits of the Wisconsin Fast Plant Inherited in a Predictable Pattern?

   a. I agree with this statement.
   b. I disagree with this statement.

   Explain your answer, using examples from your investigation about Mendelian genetics.

5. Scientists often attempt to identify patterns in nature. Explain why the identification of patterns is useful in science, using an example from your investigation about Mendelian genetics.

6. Scientists often attempt to identify the underlying cause for the observations they make. Explain why the identification of underlying causes is so important in science, using an example from your investigation about Mendelian genetics.

7. Scientists often need to look for proportional relationships. Explain what a proportional relationship is and why these relationships are important, using an example from your investigation about Mendelian genetics.

LAB 17

Lab 17. Chromosomes and Karyotypes: How Do Two Physically Healthy Parents Produce a Child With Down Syndrome and a Second Child With Cri Du Chat Syndrome?

Lab Handout

Introduction

Mendel's model of inheritance is the basis for modern genetics. This important model can be broken down into four main ideas. First, and foremost, the fundamental unit of inheritance is the gene and alternative versions of a gene (alleles) account for the variation in inheritable characters. Second, an organism inherits two alleles for each character, one from each parent. Third, if the two alleles differ, then one is fully expressed and determines the nature of the specific trait (this version of the gene is called the dominant allele) while the other one has no noticeable effect (this version of the gene is called the recessive allele). Fourth, the two alleles for each character segregate (or separate) during gamete production. Therefore, an egg or a sperm cell only gets one of the two alleles that are present in the somatic cells of the organism. This idea is known as the law of segregation.

It was brilliant (or lucky) that Mendel chose plant traits that turned out to have a relatively simple genetic basis. Each trait that he studied is determined by only one gene, and each of these genes only consists of two alleles. These conditions, however, are not met by all inheritable traits. The relationship between traits and genes is not always a simple one. In this investigation, you will use what you know about the relationship between traits and genes to explain how two children from the same family inherited two different genetic disorders.

The first child is Emily. She was born with Down syndrome. Children with Down syndrome have developmental delays, a characteristic facial appearance, and weak muscle tone. In addition, these children have an increased risk of heart defects, digestive problems such as gastroesophageal reflux, and hearing loss. The second child is Andy, Emily's younger brother. He was born with cri du chat syndrome. Children with cri du chat syndrome have severe physical and mental developmental delays, distinctive facial features, a small head (microcephaly), a low birth weight, and weak muscle tone (hypotonia).

Christopher and Jill Miller are the parents of Emily and Andy and have been married for 15 years. Although the Millers were in their early forties when they had their first child, both of them were in excellent health. They both eat a well-balanced diet and exercise on a regular basis, and they do not smoke. The Millers therefore want to know why their daughter was born with Down syndrome and their son was born with cri du chat syndrome. Here are three potential explanations:

Chromosomes and Karyotypes
How Do Two Physically Healthy Parents Produce a Child With Down Syndrome and a Second Child With Cri Du Chat Syndrome?

1. Down syndrome and cri du chat syndrome are both recessive genetic disorders. Christopher and Jill Miller each carried a recessive allele for these syndromes, and they each passed it down to their children.

2. Down syndrome and cri du chat syndrome are both caused by a chromosomal abnormality. Either the sperm cell from Christopher Miller or the egg from Jill Miller had a damaged, missing, or additional chromosome.

3. Down syndrome and cri du chat syndrome are both caused by toxins in the environment that alter genes. The children were exposed to these toxins before they were born.

Your Task
Determine which one of these explanations is most valid or acceptable.

The guiding question for this investigation is, **How do two physically healthy parents produce a child with Down syndrome and a second child with cri du chat syndrome?**

Materials
You may use any of the following materials during your investigation:

- Karyotype for Jill Miller
- Chromosome smear for Christopher Miller
- Chromosome smear for Emily Miller (born with Down syndrome)
- Chromosome smear for Andy Miller (born with cri du chat syndrome)
- 1–3 Karyotype placement grids
- Miller family pedigree

Safety Precautions

1. Wash hands with soap and water after completing this lab.
2. Follow all normal lab safety rules.

Getting Started
Unlike diseases that are transmitted from person to person, such as the flu or strep throat, people are born with cri du chat syndrome or Down syndrome. These syndromes therefore may have a genetic basis. One way to determine the underlying cause of a syndrome with a genetic basis is to produce a karyotype and then look for chromosomal abnormalities that may explain it.

A lab technician can create a karyotype by collecting a sample of cells from an individual. The sample of cells is then stained a dye that makes the chromosomes easier to see (see the figure on p. 144). Next, the chromosomes are photographed using a microscope

LAB 17

Chromosomes in a cell

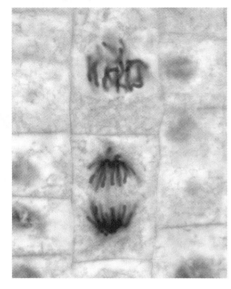

camera. The pictures of the chromosomes are organized onto a grid by size, shape, and banding pattern. Medical professionals can then use the karyotype to look for chromosomal abnormalities such as a missing chromosome or the presence of too many chromosomes. A chromosomal abnormality can also be found on a single chromosome; for example, a chromosome might be shorter or longer than it should.

To create a karyotype for Christopher Miller and the two children, you will need to sort images of chromosomes taken from their cells according to length, pair any matching sets of chromosomes, and place them onto a grid. The final product is a karyotype (a picture of an individual's chromosomes). Your teacher will provide a karyotype from Jill Miller so you can see what a normal female karyotype looks like.

Your teacher will also provide you with a pedigree for the Miller family. This pedigree will provide you with important information about the extended Miller family. It will also show the members of the Miller family that were born with either Down syndrome or cri du chat syndrome. You can use the pedigree to determine if a recessive gene could have caused one or both of these syndromes.

Investigation Proposal Required?   ☐ Yes   ☐ No

Connections to Crosscutting Concepts and to the Nature of Science and the Nature of Scientific Inquiry

As you work through your investigation, be sure to think about

- the importance of identifying patterns,
- how scientists attempt to uncover causal mechanisms,
- how structure is related to function in living things,
- how the work of scientists is influenced by social and culural values, and
- the different methods that scientists can use to answer a research question.

Argumentation Session

Once your group has finished collecting and analyzing your data, prepare a whiteboard that you can use to share your initial argument. Your whiteboard should include all the information shown in the figure on the opposite page.

To share your argument with others, we will be using a round-robin format. This means that one member of your group will stay at your lab station to share your group's argu-

Chromosomes and Karyotypes
How Do Two Physically Healthy Parents Produce a Child With Down Syndrome and a Second Child With Cri Du Chat Syndrome?

ment while the other members of your group go to the other lab stations one at a time to listen to and critique the arguments developed by your classmates.

The goal of the argumentation session is not to convince others that your argument is the best one; rather, the goal is to identify errors or instances of faulty reasoning in the arguments so these mistakes can be fixed. You will therefore need to evaluate the content of the claim, the quality of the evidence used to support the claim, and the strength of the justification of the evidence included in each argument that you see. In order to critique an argument, you will need more information than what is included on the whiteboard. You might, therefore, need to ask the presenter one or more follow-up questions, such as:

Argument presentation on a whiteboard

The Guiding Question:	
Our Claim:	
Our Evidence:	Our Justification of the Evidence:

- Is that the only way to interpret the results of your analysis? How do you know that your interpretation of your analysis is appropriate?
- Why did your group decide to present your evidence in that manner?
- Why did your group abandon the other alternative explanations?
- How confident are you that your claim is valid? What could you do to increase your confidence?

Once the argumentation session is complete, you will have a chance to meet with your group and revise your original argument. Your group might need to gather more data or design a way to test one or more alternative claims as part of this process. Remember, your goal at this stage of the investigation is to develop the most valid or acceptable answer to the research question!

Report

Once you have completed your research, you will need to prepare an investigation report that consists of three sections that provide answers to the following questions:

1. What question were you trying to answer and why?
2. What did you do during your investigation and why did you conduct your investigation in this way?
3. What is your argument?

LAB 17

Your report should answer these questions in two pages or less. This report must be typed, and any diagrams, figures, or tables should be embedded into the document. Be sure to write in a persuasive style; you are trying to convince others that your claim is acceptable or valid!

Miller family pedigree

Lab 17. Chromosomes and Karyotypes: How Do Two Physically Healthy Parents Produce a Child With Down Syndrome and a Second Child With Cri Du Chat Syndrome?

Checkout Questions

Use the picture below to answer questions 1–3.

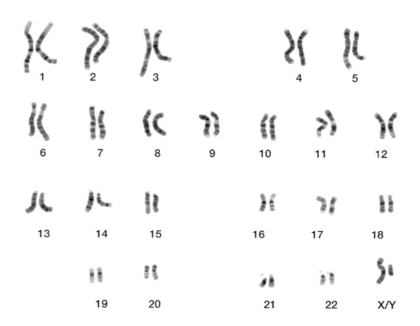

1. Is this karyotype from a male or a female?

   a. Male
   b. Female

   How do you know?

LAB 17

2. Does this person have Down syndrome?

    a. Yes
    b. No

    How do you know?

3. Does this person have cri du chat syndrome?

    a. Yes
    b. No

    How do you know?

4. Social issues and cultural values can influence the work of scientists.

    a. I agree with this statement.
    b. I disagree with this statement.

    Explain your answer, using examples from your investigation about chromosomes and karyotypes.

Chromosomes and Karyotypes
How Do Two Physically Healthy Parents Produce a Child With Down Syndrome and a Second Child With Cri Du Chat Syndrome?

5. All scientific investigations follow the same step-by-step procedure.

    a. I agree with this statement.
    b. I disagree with this statement.

    Explain your answer, using information from your investigation about chromosomes and karyotypes.

6. Scientists often attempt to identify patterns in nature. Explain why the identification of patterns is useful in science, using an example from your investigation about chromosomes and karyotypes.

7. An important goal of science is to undercover the underlying cause of observations. Explain why the identification of underlying causes is so important in science, using an example from your investigation about chromosomes and karyotypes.

8. Structure and function are related in living things. Explain what this means and why this is an important concept to keep in mind during an investigation, using an example from your investigation about chromosomes and karyotypes.

LAB 18

Lab 18. DNA Structure: What Is the Structure of DNA?

Lab Handout

Introduction

We know that genes are made of DNA because scientists were able to demonstrate that DNA and proteins are found in the nucleus of cells, and, more importantly, that DNA (and not protein) is able to transform the traits of organisms. Oswald Avery, Colin MacLeod, and Maclyn McCarty made this discovery in 1944. Their research showed that it is possible to transform harmless bacteria into infectious ones with pure DNA. They also provided further support for their claim by demonstrating that it is possible to prevent this "'transformation" with a DNA-digesting enzyme called DNase.

However, knowing that genes are made of DNA and that DNA is able to store the genetic information of an individual is a little like having a parts list to a 747 jumbo jet. It tells what is important, but it tells you little about how it works. To figure out how DNA works—that is, how it is able to store genetic information—scientists had to figure out its structure. In this investigation, you will duplicate the work of the two scientists who first figured out the structure of DNA—James Watson and Francis Crick.

Your Task

Use the available data to develop a model that explains the structure of DNA.

The guiding question of this investigation is, **What is the structure of DNA?**

Materials

You may use any of the following materials during your investigation:

- Pop beads (DNA kit)
- Fact sheet about DNA

Safety Precautions

1. Safety goggles or glasses are required for this lab.
2. Wash hands with soap and water after completing this lab.
3. Follow all normal lab safety rules.

DNA Structure
What Is the Structure of DNA?

Getting Started

To answer the guiding question, you will need to develop a model for the structure of DNA. In science, models are explanations for how things work or how they are structured. Scientists often need to develop models to explain a complex phenomenon or to understand the structure of things that are too small to see (such as the structure of an atom or the structure of a molecule of DNA). Scientists use drawings, graphs, equations, three-dimensional representations, or words to communicate their models to others, but scientists only use these physical objects as a way to illustrate the major components of the model.

You will need to create a three-dimensional representation of your model for the structure of DNA using pop beads. Remember that more than one model may be an acceptable explanation for the same phenomenon. It is not always possible to exclude all but one model—and also not always desirable. For example, physicists think about light as a wave and as a particle, and each model of light's behavior is used to think about and account for phenomena differently.

Investigation Proposal Required?   ☐ Yes   ☐ No

Connections to Crosscutting Concepts and to the Nature of Science and Scientific Inquiry

As you work through your investigation, be sure to think about

- the importance of identifying patterns;
- the importance of examining proportional relationships;
- how the way an object is shaped or structured determines many of its properties or functions;
- how science, as a body of knowledge, changes over time; and
- the different methods that scientists can use to answer a research question.

Argumentation Session

Once your group has finished collecting and analyzing your data, prepare a whiteboard that you can use to share your initial argument. Your whiteboard should include all the information shown in the figure to the right.

To share your argument with others, we will be using a round-robin format. This means that one member of your group will stay at your lab station to share your group's argument while the other

Argument presentation on a whiteboard

The Guiding Question:	
Our Claim:	
Our Evidence:	Our Justification of the Evidence:

members of your group go to the other lab stations one at a time to listen to and critique the arguments developed by your classmates.

The goal of the argumentation session is not to convince others that your argument is the best one; rather, the goal is to identify errors or instances of faulty reasoning in the arguments so these mistakes can be fixed. You will therefore need to evaluate the content of the claim, the quality of the evidence used to support the claim, and the strength of the justification of the evidence included in each argument that you see. In order to critique an argument, you will need more information than what is included on the whiteboard. You might, therefore, need to ask the presenter one or more follow-up questions, such as:

- What did you do to develop your model?
- Is that the only way to interpret the results of your analysis? How do you know that your interpretation of your analysis is appropriate?
- Why did your group decide to present your evidence in that manner?
- What other models did your group discuss before you decided on that one? Why did your group abandon those alternative ideas?
- How confident are you that your model is valid? What could you do to increase your confidence?

Once the argumentation session is complete, you will have a chance to meet with your group and revise your original argument. Your group might need to gather more data or design a way to test one or more alternative claims as part of this process. Remember, your goal at this stage of the investigation is to develop the most valid or acceptable answer to the research question!

Report

Once you have completed your research, you will need to prepare an investigation report that consists of three sections that provide answers to the following questions:

1. What question were you trying to answer and why?
2. What did you do during your investigation and why did you conduct your investigation in this way?
3. What is your argument?

Your report should answer these questions in two pages or less. This report must be typed, and any diagrams, figures, or tables should be embedded into the document. Be sure to write in a persuasive style; you are trying to convince others that your claim is acceptable or valid.

DNA Structure
What Is the Structure of DNA?

DNA Fact Sheet

1. DNA is a very long molecule composed of smaller molecules called subunits. You can use the different colored beads to represent the different subunits in your physical representation of DNA.

2. DNA is composed of six different subunits (or smaller molecules):

   Guanine (a base)   Phosphate group   Thymine (a base)

   Deoxyribose (a sugar)   Adenine (a base)   Cytosine (a base)

3. DNA consists of two chains that are bonded (connected) together. A subunit from one strand bonds to a subunit on the other.

4. The diameter of DNA is the same along its entire length (exactly four molecules or subunits wide). Rosalind Franklin made this discovery in 1952 by using x-ray diffraction (see the figure below).

X-rays show that DNA has the same diameter along its entire length.

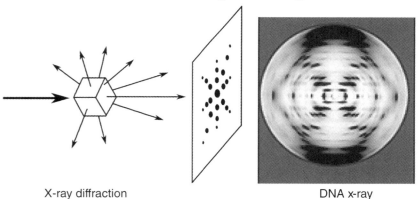

X-ray diffraction   DNA x-ray

5. A sugar subunit can only bind with two other subunits: a base subunit and a phosphate group subunit.

6. A base subunit can only bind with two other subunits: a sugar subunit and a base subunit.

7. A phosphate group subunit can only bind with a sugar subunit.

8. In 1950, biochemist Erwin Chargaff examined the proportion of adenine (A), thymine (T), guanine (G), and cytosine (C) molecules in DNA from different types of organisms. His findings, which are shown below, were so important that it led

LAB 18

to a fundamental principle about the relative proportion of bases found in the DNA of all organisms; this principle is now known as Chargaff's rules.

Relative proportions (%) of bases in DNA				
Organism	A	T	G	C
Human	30.9	29.4	19.9	19.8
Chicken	28.8	29.2	20.5	21.5
Grasshopper	29.3	29.3	20.5	20.7
Sea urchin	32.8	32.1	17.7	17.3
Wheat	27.3	27.1	22.7	22.8
Yeast	31.3	32.9	18.7	17.1
E. coli	24.7	23.6	26.0	25.7

Lab 18. DNA Structure: What Is the Structure of DNA?

Checkout Questions

1. In the space below, draw a picture that illustrates the structure of DNA. Be sure to include and label the four different bases, the phosphate group, and the deoxyribose.

2. All scientific knowledge, including the findings from studies that have been published in peer-reviewed journals, is subject to ongoing testing and revision.

   a. I agree with this statement.
   b. I disagree with this statement.

   Explain your answer, using information from your investigation about DNA structure.

3. When conducting a new investigation, scientists can use data previously gathered by other scientists.

   a. I agree with this statement.
   b. I disagree with this statement.

   Explain your answer, using examples from your investigation about DNA structure.

LAB 18

4. Scientists often attempt to identify patterns in nature. Explain why the identification of patterns is useful in science, using an example from your investigation about DNA structure.

5. Scientists often need to look for proportional relationships when analyzing data. Explain why it is often useful to look for these relationships in science, using an example from your investigation about DNA structure.

6. Scientists often attempt to determine the structure of molecules that are too small to see. Explain why this is important for scientists to do, using an example from your investigation about DNA structure.

Lab 19. Meiosis: How Does the Process of Meiosis Reduce the Number of Chromosomes in Reproductive Cells?

Lab Handout

Introduction

Sexual reproduction is a process that creates a new organism by combining the genetic material of two organisms. There are two main steps in sexual reproduction: (1) the production of reproductive cells and (2) fertilization. Fertilization is the fusion of two reproductive cells to form a new individual. During fertilization, chromosomes from the male and female combine to form homologous pairs (see the figure below). The number of chromosomes donated from the male and female are equal, and offspring have the same number of chromosomes as each of the parents.

A human karyotype depicting 23 homologous pairs of chromosomes

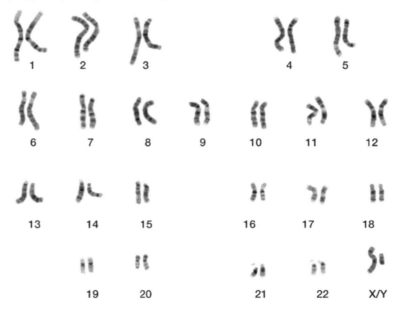

If the reproductive cell from a male and the reproductive cell from a female each donate the same number of chromosomes that are found in a typical cell to the new embryo, then the chromosome number of the species would double with each generation. Yet that doesn't happen; the chromosome number within a species stays constant from one generation to the next. Therefore, a mechanism that reduces the number of chromosomes found

LAB 19

in reproductive cells is needed to prevent the chromosome number from doubling as the result of fertilization. This mechanism is called meiosis.

It took many years of research and contributions from several different scientists to uncover what happens inside a cell during the complex process of meiosis. The German biologist Oscar Hertwig made the first major contribution in 1876, when he documented the stages of meiosis by examining the formation of eggs in sea urchins. The Belgian zoologist Edouard Van Beneden made the next major contribution in 1883. He was the first to describe the basic behavior of chromosomes during meiosis by studying the formation of eggs in an intestinal worm (Ascaris). Finally, the German biologist August Weismann highlighted the potential significance of meiosis for reproduction and inheritance in 1890. Weismann was the first one to publish an article that suggested that meiosis could half the number of chromosomes in reproductive cells and, as a result, keep the chromosome number within a species constant from one generation to the next. In this investigation, you will attempt to build on the work of these scientists by developing a model that explains how this type of cell division results in the production of reproductive cells that contain halve the number of chromosomes that are found in the other cells of that organism.

Your Task

Meiosis is the process in which chromosomes are duplicated and then separated into four reproductive cells that have exactly half the number of chromosomes of the original organism. In addition, this process ensures that there are no pairs of chromosomes found in the reproductive cells. In other words, there is only one copy of each chromosome in reproductive cells instead of two. Then during fertilization, a reproductive cell from a male (i.e., a sperm) and a reproductive cell from a female (i.e., an egg) will fuse to form an embryo that has the same number of chromosomes as the original organism. This process happens in all animals, plants, and fungi that reproduce sexually. Your goal is to develop a model that explains what happens to the chromosomes within a cell during each stage of meiosis.

The guiding question of this investigation is, **How does the process of meiosis reduce the number of chromosomes in reproductive cells?**

Materials

You may use any of the following materials:

- 8 Pop bead chromosomes (and extra pop beads if needed)
- Images of the stages of meiosis

Safety Precautions

1. Safety goggles or glasses are required for this lab.

Meiosis
How Does the Process of Meiosis Reduce the Number of Chromosomes in Reproductive Cells?

2. Wash hands with soap and water after completing this lab.
3. Follow all normal lab safety rules.

Getting Started

To answer the guiding question, you will need to develop a model that explains the process of meiosis. Your first step is to learn more about what a cell looks like as it goes through meiosis. You will therefore be given a series of pictures that show the different stages of meiosis as seen through a microscope. Your next step will be to figure out (a) the correct order of stages and (b) what you think may be going on during each stage. Use what you know about the stages of mitosis and how cells divided during mitosis to accomplish this task. From there, you can use pop bead chromosomes (see the figure to the right) to attempt to make sense of what is happening with the individual chromosomes during each stage of meiosis.

Pop bead chromosomes

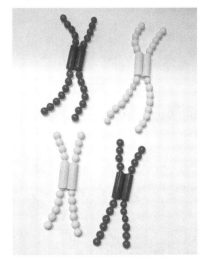

Your model, once complete, should be able to explain (a) what happens to the chromosomes inside a cell as it goes though meiosis, (b) why reproductive cells have half the number of chromosomes of the individuals who produce them, and (c) why there are no pairs of chromosomes in reproductive cells. To be valid, your model must be able to explain all three of these issues.

Investigation Proposal Required? ☐ Yes ☐ No

Connections to Crosscutting Concepts and to the Nature of Science and Scientific Inquiry

As you work through your investigation, be sure to think about

- the importance of identifying and explaining patterns,
- how scientists develop and use models,
- the difference between theories and laws in science, and
- the role of creativity and imagination in science.

Argumentation Session

Once your group has finished collecting and analyzing your data, prepare a whiteboard that you can use to share your initial argument. Your whiteboard should include all the information shown in the figure on p. 160.

LAB 19

Argument presentation on a whiteboard

The Guiding Question:	
Our Claim:	
Our Evidence:	Our Justification of the Evidence:

To share your argument with others, we will be using a round-robin format. This means that one member of your group will stay at your lab station to share your group's argument while the other members of your group go to the other lab stations one at a time to listen to and critique the arguments developed by your classmates.

The goal of the argumentation session is not to convince others that your argument is the best one; rather, the goal is to identify errors or instances of faulty reasoning in the arguments so these mistakes can be fixed. You will therefore need to evaluate the content of the claim, the quality of the evidence used to support the claim, and the strength of the justification of the evidence included in each argument that you see. In order to critique an argument, you will need more information than what is included on the whiteboard. You might, therefore, need to ask the presenter one or more follow-up questions, such as:

- Is that the only way to interpret the results of your analysis? How do you know that your interpretation of your analysis is appropriate?
- Why did your group decide to present your evidence in that manner?
- What other models did your group discuss before you decided on that one? Why did your group abandon those alternative ideas?
- How confident are you that your model is valid? What could you do to increase your confidence?

Once the argumentation session is complete, you will have a chance to meet with your group and revise your original argument. Your group might need to gather more data or design a way to test one or more alternative claims as part of this process. Remember, your goal at this stage of the investigation is to develop the most valid or acceptable answer to the research question!

Report

Once you have completed your research, you will need to prepare an investigation report that consists of three sections that provide answers to the following questions:

1. What question were you trying to answer and why?
2. What did you do during your investigation and why did you conduct your investigation in this way?
3. What is your argument?

Meiosis
How Does the Process of Meiosis Reduce the Number of Chromosomes in Reproductive Cells?

Your report should answer these questions in two pages or less. This report must be typed, and any diagrams, figures, or tables should be embedded into the document. Be sure to write in a persuasive style; you are trying to convince others that your claim is acceptable or valid!

LAB 19

Lab 19. Meiosis: How Does the Process of Meiosis Reduce the Number of Chromosomes in Reproductive Cells?

Checkout Questions

1. Describe the process and products of meiosis.

2. The idea that genes are found on chromosomes and that there is the process of meiosis are theories and not laws.

    a. I agree with this statement.
    b. I disagree with this statement.

    Explain your answer, using information from your investigation about meiosis.

Meiosis
How Does the Process of Meiosis Reduce the Number of Chromosomes in Reproductive Cells?

3. Creativity and imagination are an important part of this investigation.

    a. I agree with this statement.
    b. I disagree with this statement.

    Explain your answer, using examples from your investigation about meiosis.

4. Scientists often attempt to identify and explain the patterns in nature. Explain why the identification of patterns is useful in science, using an example from your investigation about meiosis.

5. Scientists often attempt to develop a conceptual model to explain complex phenomena. Explain what a model is and why models are useful in science, using an example from your investigation about meiosis.

Lab 20. Inheritance of Blood Type: Are All of Mr. Johnson's Children His Biological Offspring?

Lab Handout

Introduction

Karl Landsteiner identified the ABO blood group in 1901. The ABO blood group includes four types of blood (A, B, AB, and O). The differences in blood types are due to the presence or absence of certain types of antigens and antibodies. Antigens are molecules that are located on the surface of the red blood cells (RBCs), and antibodies are molecules that are located in the blood plasma. Individuals have different types and combinations of these molecules. The figure below shows which antigens and antibodies are associated with each blood type in the ABO blood group.

Blood types and red blood cell surface antigens

	Type A	Type B	Type AB	Type O
Red blood cell type	A	B	AB	O
Antibodies in Plasma	Anti-B	Anti-A	None	Anti-A and Anti-B
Antigens in Red Blood Cell	A antigen	B antigen	A and B antigens	None

A single gene that consists of three different versions (or alleles) determines the four blood types in the ABO group. Allele A codes for the synthesis of RBCs that have the type A antigens on their surface. Allele B codes for the synthesis of RBCs that have the type B antigens on their surface, and allele O codes for RBCs that lack surface antigens. The A and B alleles are codominant to each other, and both the A and B alleles are dominant over the O allele. Although there are three different alleles associated with the ABO blood group gene, each individual only inherits two copies of it. One copy of the gene comes from the mother and one copy of the gene comes from the father. The ABO blood type therefore follows the multiple allele model of inheritance.

Although blood type is an inherited trait, the U.S. judicial system does not recognize ABO blood typing as an acceptable way to determine paternity because many individuals can have the same blood type. In the United States, for example, approximately 44% of the population has type O blood, 42% has type A blood, 10% has type B blood, and 4% has type AB blood. ABO blood-typing, however, can be used to exclude a man from being a child's father. Therefore, it is sometimes useful to conduct a quick and inexpensive test for ABO blood type to determine if further testing using a DNA analysis is warranted.

Your Task

Mr. and Mrs. Johnson have been married for eight years. During this time, Mrs. Johnson has had three children. Recently Mr. Johnson found out that Mrs. Johnson has been secretly dating another man, Mr. Wilson, throughout their marriage. Mr. Johnson now questions if he is truly the biological father of the three children. Your goal is to use what you know about the inheritance of ABO blood types to determine if Mr. Johnson can be excluded as the father of any of Mrs. Johnson's children.

The guiding question of this investigation is, **Are all of Mr. Johnson's children his biological offspring?**

Materials

You may use any of the following materials during your investigation:

- Type A blood sample
- Type B blood sample
- Type AB blood sample
- Type O blood sample
- Blood sample from Mr. Wilson
- Blood sample from Mr. Johnson
- Blood sample from Mrs. Johnson
- Blood sample from child 1
- Blood sample from child 2
- Blood sample from child 3
- Anti-A serum
- Anti-B serum
- 6 blood-typing slides
- Toothpicks

Safety Precautions

1. Safety goggles, gloves, and aprons are required for this lab.
2. Under no circumstances is human or animal blood to be used or tested. Only use commercially prepared simulated blood products.
3. Wash hands with soap and water after completing the lab.
4. Follow all normal lab safety rules.

LAB 20

Getting Started

To test a person's blood type, you can mix a sample of blood with an antiserum that has high levels of anti-A or anti-B antibodies. The simple test is performed as follows:

1. Add two drops of a blood sample to well A and to well B of a blood-typing slide.
2. Add two drops of the appropriate antiserum to each of the samples.
3. Stir each sample for 20 seconds with a toothpick.

If the blood cells have the appropriate antigens on their surface, agglutination (clumping of the blood) will occur. For example, if anti-A serum is added to a sample of blood and agglutination occurs, that means the blood contains cells that have the type A antigens on their surface. The figure below illustrates the reaction of each antiserum with each blood type. Be sure to test known samples first before the unknown samples to see what the agglutination reactions look like.

Investigation Proposal Required?   ☐ Yes   ☐ No

Connections to Crosscutting Concepts and to the Nature of Science and Scientific Inquiry

As you work through your investigation, be sure to think about

- how scientists develop and use explanatory models to make sense of their observations,

Reaction of different blood types with antiserum

Antiserum	Reaction when blood is mixed with antiserum			
	Type A	Type B	Type AB	Type O
Anti-B	(no agglutination)	(agglutination)	(agglutination)	(no agglutination)
Anti-A	(agglutination)	(no agglutination)	(agglutination)	(no agglutination)

- the relationship between structure and function,
- the relationship between observations and inferences in science, and
- the difference between data and evidence.

Argumentation Session

Once your group has finished collecting and analyzing your data, prepare a whiteboard that you can use to share your initial argument. Your whiteboard should include all the information shown in the figure to the right.

To share your argument with others, we will be using a round-robin format. This means that one member of your group will stay at your lab station to share your group's argument while the other members of your group go to the other lab stations one at a time to listen to and critique the arguments developed by your classmates.

Argument presentation on a whiteboard

The Guiding Question:	
Our Claim:	
Our Evidence:	Our Justification of the Evidence:

The goal of the argumentation session is not to convince others that your argument is the best one; rather, the goal is to identify errors or instances of faulty reasoning in the arguments so these mistakes can be fixed. You will therefore need to evaluate the content of the claim, the quality of the evidence used to support the claim, and the strength of the justification of the evidence included in each argument that you see. In order to critique an argument, you will need more information than what is included on the whiteboard. You might, therefore, need to ask the presenter one or more follow-up questions, such as:

- How did you collect your data? Why did you use that method? Why did you collect those data?
- What did you do to make sure the data you collected are reliable? What did you do to decrease measurement error?
- What did you do to analyze your data? Why did you decide to do it that way?
- Is that the only way to interpret the results of your analysis? How do you know that your interpretation of your analysis is appropriate?
- Why did your group decide to present your evidence in that manner?
- What other claims did your group discuss before you decided on that one? Why did your group abandon those alternative ideas?
- How confident are you that your claim is valid? What could you do to increase your confidence?

LAB 20

Once the argumentation session is complete, you will have a chance to meet with your group and revise your original argument. Your group might need to gather more data or design a way to test one or more alternative claims as part of this process. Remember, your goal at this stage of the investigation is to develop the most valid or acceptable answer to the research question!

Report

Once you have completed your research, you will need to prepare an investigation report that consists of three sections that provide answers to the following questions:

1. What question were you trying to answer and why?
2. What did you do during your investigation and why did you conduct your investigation in this way?
3. What is your argument?

Your report should answer these questions in two pages or less. This report must be typed, and any diagrams, figures, or tables should be embedded into the document. Be sure to write in a persuasive style; you are trying to convince others that your claim is acceptable or valid!

Lab 20. Inheritance of Blood Type: Are All of Mr. Johnson's Children His Biological Offspring?

Checkout Questions

1. Two tall pea plants are cross-fertilized and produce four offspring. Three of the four offspring are tall plants, and one of the offspring is short.

Parents:	Tall Plant	×	Tall Plant	
Offspring:	Tall Plant	Tall Plant	Tall Plant	Short Plant

   Use Mendel's model of inheritance to explain how two tall plants can produce tall plants and short plants.

2. In science the terms *data* and *evidence* have the same meaning.

   a. I agree with this statement.
   b. I disagree with this statement.

   Explain your answer, using examples from your investigation about inheritance of blood type.

LAB 20

3. Scientists will always make the same observations and inferences during an investigation.

    a. I agree with this statement.
    b. I disagree with this statement.

    Explain your answer, using information from your investigation about inheritance of blood type.

4. Scientists often use models to explain a complex phenomenon. Explain why models are useful in science, using an example from your investigation about inheritance of blood type.

5. Structure and function are related in nature. Explain how this principle influenced how you collected data and how you made sense of the data you collected during your investigation about inheritance of blood type.

Lab 21. Models of Inheritance: Which Model of Inheritance Best Explains How a Specific Trait Is Inherited in Fruit Flies?

Lab Handout

Introduction

The principles of Mendelian genetics encompass several different models of inheritance. These models include dominant-recessive, incomplete dominance, codominance, multiple allele, and sex-linked. All these models share two key ideas. First, inherited genes determine specific traits and alternative versions of a gene (called alleles) are responsible for the variation we see in traits. Second, an organism inherits two alleles for each trait, one from each parent. What makes these models different from each other is in the nature of the interactions that occur between alleles, the number of types of alleles that are associated with a gene, or which chromosome carries the gene for a trait.

The dominant-recessive model of inheritance suggests that when an individual inherits two alleles and the two alleles differ, then one is fully expressed and determines the nature of a trait (this version of the gene is called the dominant allele) while the other one has no noticeable effect (this version of the gene is called the recessive allele). The incomplete dominance model of inheritance suggests that the interaction that occurs between two different alleles results in a hybrid with an appearance somewhere between the phenotypes of the two parental varieties. The codominance model of inheritance is similar to the incomplete dominance model, but in the codominance model both alleles affect the phenotype of the individual in separate and distinguishable ways. The multiple allele model of inheritance simply means that there are more than two versions of a gene for a given trait within a population, and each allele can be either dominant, recessive, incomplete dominant, or codominant to the other alleles. Finally, the sex-linked model of inheritance only applies to genes located on the sex chromosomes. Females and males differ in the number of genes they inherit when the gene is found on the sex chromosome; one gender inherits two copies of the gene and the other gender inherits only one.

In this investigation, you will use what you have learned about these different models of inheritance to determine how fruit flies inherit a specific trait. Fruit flies are very common. Most fruit flies have six legs, two wings, and two antennae (see the figure on p. 172). Most fruit flies also have an orange-yellow body and red eyes. Scientists call flies with these traits the "wild type." Every once in a while, however, you might see a fruit fly with body and/or eyes that are different colors, such as black or yellow body and sepia (brown) or white eyes.

LAB 21

Male and female fruit flies

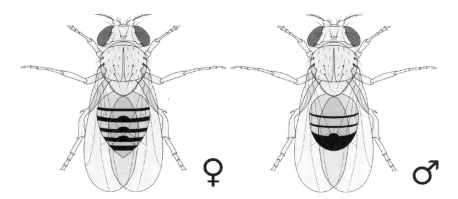

Your Task

Pick two fruit fly traits (e.g., eye color, body color, wing shape). Then you will need to determine which model of inheritance (dominant-recessive, incomplete dominance, codominance, multiple allele, or sex-linked) best explains the inheritance pattern of these two specific traits in fruit flies. To accomplish this goal, you will use an online simulation that allows you to "order" fruit flies with specific traits from a supply company and then "breed" them to see how a trait is passed down from parent to offspring.

The guiding question of this investigation is, **Which model of inheritance best explains how a specific trait is inherited in fruit flies?**

Materials

You will use an online simulation called *Drosophila* to conduct your investigation. You can access the simulation by going to the following website: *www.sciencecourseware.org/vcise/drosophila* .

Safety Precautions

1. Use caution when working with electrical equipment. Keep away from water sources in that they can cause shorts, fires, and shock hazards. Use only GFI-protected circuits.

2. Wash hands with soap and water after completing this lab.

3. Follow all normal lab safety rules.

Getting Started

Your teacher will show you how to use the Drosophila online simulation and the types of traits you will be able to investigate before you begin designing your investigation.

Models of Inheritance
Which Model of Inheritance Best Explains How a Specific Trait Is Inherited in Fruit Flies?

To answer the guiding question, you will need to determine what type of data you will need to collect using the online simulation, how you will collect these data, and how you will analyze the data. To determine *what type of data you will need to collect*, think about the following questions:

- What types of flies will you need to work with during your investigation (e.g., males or females, flies with a specific eye color, flies with a specific wing shape)?
- What type of measurements or observations will you need to record during your investigation?
- How will you determine which model of inheritance is the best explanation for a particular trait?

To determine *how you will collect your data*, think about the following questions:

- How many times will you need to breed the flies?
- How many generations of flies will you need to follow?
- How often will you collect data and when will you do it?
- How will you keep track of the data you collect and how will you organize the data?

To determine *how you will analyze your data*, think about the following questions:

- How will you determine if the results of your cross tests match your predictions? (Hint: Your teacher will show you how to use a statistical method called a chi-square test to help determine if your observations match your prediction once you have collected all your data.)
- What type of graph could you create to help make sense of your data?

Investigation Proposal Required?   ☐ Yes   ☐ No

Connections to Crosscutting Concepts and to the Nature of Science and Scientific Inquiry

As you work through your investigation, be sure to think about

- the importance of uncovering causes for patterns observed in nature,
- how scientists develop and use explanatory models to make sense of their observations, and
- the nature of theories and laws in science.

LAB 21

Argumentation Session

Once your group has finished collecting and analyzing your data, prepare a whiteboard that you can use to share your initial argument. Your whiteboard should include all the information shown in the figure below.

Argument presentation on a whiteboard

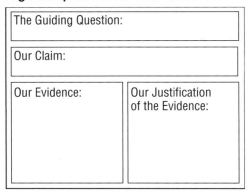

To share your argument with others, we will be using a round-robin format. This means that one member of your group will stay at your lab station to share your group's argument while the other members of your group go to the other lab stations one at a time to listen to and critique the arguments developed by your classmates.

The goal of the argumentation session is not to convince others that your argument is the best one; rather, the goal is to identify errors or instances of faulty reasoning in the arguments so these mistakes can be fixed. You will therefore need to evaluate the content of the claim, the quality of the evidence used to support the claim, and the strength of the justification of the evidence included in each argument that you see. In order to critique an argument, you will need more information than what is included on the whiteboard. You might, therefore, need to ask the presenter one or more follow-up questions, such as:

- How did you use the simulation to collect your data?
- What did you do to analyze your data? Why did you decide to do it that way? Did you check your calculations?
- Is that the only way to interpret the results of your analysis? How do you know that your interpretation of your analysis is appropriate?
- Why did your group decide to present your evidence in that manner?
- What other claims did your group discuss before you decided on that one? Why did your group abandon those alternative ideas?
- How confident are you that your claim is valid? What could you do to increase your confidence?

Once the argumentation session is complete, you will have a chance to meet with your group and revise your original argument. Your group might need to gather more data or design a way to test one or more alternative claims as part of this process. Remember, your goal at this stage of the investigation is to develop the most valid or acceptable answer to the research question!

Models of Inheritance
Which Model of Inheritance Best Explains How a Specific Trait Is Inherited in Fruit Flies?

Report

Once you have completed your research, you will need to prepare an investigation report that consists of three sections that provides answers to the following questions:

1. What question were you trying to answer and why?
2. What did you do during your investigation and why did you conduct your investigation in this way?
3. What is your argument?

Your report should answer these questions in two pages or less. This report must be typed, and any diagrams, figures, or tables should be embedded into the document. Be sure to write in a persuasive style; you are trying to convince others that your claim is acceptable or valid!

LAB 21

Lab 21. Models of Inheritance: Which Model of Inheritance Best Explains How a Specific Trait Is Inherited in Fruit Flies?

Checkout Questions

1. Color blindness is a sex-linked trait in humans. Both mother and father have normal color vision. Is it possible for their children to be color-blind?

    a. Yes
    b. No

    Explain why.

2. Scientific laws are theories that have been proven true.

    a. I agree with this statement.
    b. I disagree with this statement.

    Explain your answer, using information from your investigation about models of inheritance.

3. Mendelian genetics is based on the idea that gametes (reproductive cells) only get one of the two alleles that are present in the somatic (body) cells of the organism. Is this idea an example of a theory or a law?

    a. This idea is an example of a theory.
    b. This idea is an example of a law.

Explain your answer, using information from your investigation about models of inheritance.

4. Scientists often attempt to explain observed patterns in nature. Explain why they do this, using an example from your investigation about models of inheritance.

5. Scientists develop and or use models to explain complex phenomena. Explain what models are and why scientists view them as valuable, using an example from your investigation about models of inheritance.

SECTION 5
Life Sciences Core Idea 4:

Biological Evolution: Unity and Diversity

Lab 22. Biodiversity and the Fossil Record: How Has Biodiversity on Earth Changed Over Time?

Lab Handout

Introduction

Biodiversity refers to the variation in life forms found on Earth. Biodiversity can be measured in two different ways. The first is richness, which refers to the total number of different life forms. The second is relative abundance, which is a measure of how common each type of life form is in a given area. In terms of richness, Earth is high in biodiversity—biologists have identified approximately 1.5 million different types of life forms, and some biologists think that the actual number of different life forms on Earth is at least 7 million.

To help organize and make sense of this biodiversity, biologists use a nested classification scheme. This system (see the figure to the right) starts with species as the foundational unit of classification. A species is often defined as a group of organisms capable of interbreeding and producing fertile offspring. Each species can then be placed into a larger group called a genus, based on similarities in traits. Each genus can then be placed into a larger group called a family. Families, in turn, can be grouped together to create an order; and so on.

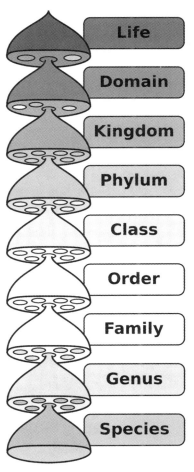

The biological classification scheme

There have been several different hypotheses offered to explain the source of all the biodiversity on Earth and the amount of biodiversity found on Earth over time. Here are three of these hypotheses:

1. All life on Earth appeared at the same time in Earth's history. As a result, biodiversity has remained the same throughout Earth's history.

2. Present-day forms of life arose from other forms of life over a considerable amount of time. As a result, biodiversity has increased throughout Earth's history.

3. All life on Earth appeared at the same time in Earth's history. However, current life forms are the survivors of one or more catastrophic events that wiped out many of the other life forms that once inhabited the Earth. As a result, biodiversity has decreased throughout Earth's history.

LAB 22

You can evaluate the merits of these three hypotheses by determining if they are consistent with what is found in the fossil record. Scientists, over many years, have collected data about the history of life on Earth. These data include the collection, classification, and dating of fossils. This information allows scientists to determine what the conditions were like on Earth in the past and when major events occurred in the history of life. It is important to note, however, that the fossil record provides only an incomplete picture of what life on Earth was like in the past. Although the fossil record is substantial, it is incomplete because life forms that are abundant, widespread, and have hard shells or skeletons are more likely to be preserved as fossils than are life forms that are rare, live in only specific locations, or have soft bodies. The fossil record, therefore, can only provide limited information about the history of life on Earth.

Your Task
Determine if the fossil record supports any of the three hypotheses listed above.

The guiding question of this investigation is, **How has biodiversity on Earth changed over time?**

Materials
You will use an Excel file called "Diversity in the Fossil Record Data," which can be found at *www.nsta.org/publications/press/extras/argument.aspx*, in your investigation.

Safety Precautions
1. Use caution when working with electrical equipment. Keep away from water sources in that they can cause shorts, fires and shock hazards. Use only GFI protected circuits.
2. Wash hands with soap and water after completing this lab.
3. Follow all normal lab safety rules.

Getting Started
The data file that you will use during this investigation comes from the Fossil Record website (*www.fossilrecord.net*). The Excel data file is a simplified version of the original Fossil Record 2 Excel file, which can be downloaded from the website. The Fossil Record 2 is a near-complete listing of the diversity of life through time, compiled at the level of the family by Mary Benton and originally published as a book in 1993. In biology, the term family refers to a taxonomic rank that falls between order and genus. The levels of classification include species, genus, family, order, class, phylum, and kingdom. For example, the Bonobo (*Pan paniscus*) is a part of the genus Pan, the family Hominidae, the order Primates,

Biodiversity and the Fossil Record
How Has Biodiversity on Earth Changed Over Time?

the class Mammalia, the phylum Chordata, and the kingdom Animalia. Many different species make up a particular family.

In the data file, there are tabs for 10 different classes of organisms (Mammalia, Reptilia, Amphibia, and so on). Each tab includes information about all the families within that class that have been found in the fossil record. The dates represent the midpoint of different geologic stages. For example, 0.001 mya is the midpoint of the Holocene stage and 0.8 mya is the midpoint of the Pleistocene stage. The number 1 in a cell indicates that a fossil belonging to that particular family has been found for that time period. A question mark (?) in a cell indicates that a fossil is likely to be found for a specific time period (because there are fossils from that family in earlier and later time periods) but has not been discovered yet. Cells that are blank indicate that there are no fossils from that family in that time period.

Investigation Proposal Required?   ☐ Yes   ☐ No

Connections to Crosscutting Concepts and to the Nature of Science and the Nature of Scientific Inquiry

As you work through your investigation, be sure to think about

- the importance of looking for patterns in nature,
- the importance of considering what is and what is not relevant at different scales of time,
- the difference between data and evidence in science, and
- the different types of methods that scientists use to answer questions.

Argumentation Session

Once your group has finished collecting and analyzing your data, prepare a whiteboard that you can use to share your initial argument. Your whiteboard should include all the information shown in the figure to the right.

Argument presentation on a whiteboard

The Guiding Question:	
Our Claim:	
Our Evidence:	Our Justification of the Evidence:

To share your argument with others, we will be using a round-robin format. This means that one member of your group will stay at your lab station to share your group's argument while the other members of your group go to the other lab stations one at a time to listen to and critique the arguments developed by your classmates.

The goal of the argumentation session is not to convince others that your argument is the best one;

LAB 22

rather, the goal is to identify errors or instances of faulty reasoning in the arguments so these mistakes can be fixed. You will therefore need to evaluate the content of the claim, the quality of the evidence used to support the claim, and the strength of the justification of the evidence included in each argument that you see. In order to critique an argument, you will need more information than what is included on the whiteboard. You might, therefore, need to ask the presenter one or more follow-up questions, such as:

- Why did you decide to focus on those data?
- What did you do to analyze your data? Why did you decide to do it that way? Did you check your calculations?
- Is that the only way to interpret the results of your analysis? How do you know that your interpretation of your analysis is appropriate?
- Why did your group decide to present your evidence in that manner?
- What other claims did your group discuss before you decided on that one? Why did your group abandon those alternative ideas?
- How confident are you that your claim is valid? What could you do to increase your confidence?

Once the argumentation session is complete, you will have a chance to meet with your group and revise your original argument. Your group might need to gather more data or design a way to test one or more alternative claims as part of this process. Remember, your goal at this stage of the investigation is to develop the most valid or acceptable answer to the research question!

Report

Once you have completed your research, you will need to prepare an investigation report that consists of three sections that provide answers to the following questions:

1. What question were you trying to answer and why?
2. What did you do during your investigation and why did you conduct your investigation in this way?
3. What is your argument?

Your report should answer these questions in two pages or less. This report must be typed, and any diagrams, figures, or tables should be embedded into the document. Be sure to write in a persuasive style; you are trying to convince others that your claim is acceptable or valid!

Lab 22. Biodiversity and the Fossil Record: How Has Biodiversity on Earth Changed Over Time?

Checkout Questions

1. What does the fossil record suggest about how biodiversity on Earth has changed over time?

2. Evidence is data that have been collected by a scientist.

    a. I agree with this statement.
    b. I disagree with this statement.

    Explain your answer, using examples from your investigation about the fossil record.

3. Using an existing data set to test several hypotheses is an example of an experiment.

    a. I agree with this statement.
    b. I disagree with this statement.

    Explain your answer, using information from your investigation about the fossil record.

LAB 22

4. Scientists often need to determine what is and what is not relevant at different time scales. Explain why this is important in science, using an example from your investigation about the fossil record.

5. Scientists often attempt to identify patterns in nature. Explain why the identification of patterns is useful in science, using an example from your investigation about the fossil record.

Lab 23. Mechanisms of Evolution: Why Will the Characteristics of a Bug Population Change in Different Ways in Response to Different Types of Predation?

Lab Handout

Introduction

The various components of an ecosystem are all connected. Plants depend on the abiotic resources of an ecosystem to produce the food they need to grow, herbivores eat these plants, and carnivores eat the herbivores. Thus, a change in the amount of abiotic resources available or a change in the size of any one of these populations of organisms can influence the size of the other populations found in that ecosystem. A drought, for example, could reduce the size of the plant population. A decrease in the size of the plant population results in less food for the herbivores. When herbivores do not have enough food to eat, the death rate of the population increases, which, in turn, results in fewer herbivores. The size of the carnivore population, as a result, begins to shrink because there is not enough food available.

In addition to influencing the size of a population, the interactions that take place between the organisms found within an ecosystem can actually change the characteristics of some populations. Some of the characteristics that can be influenced by these interactions include the ratio of males to females in a population or the ratio of juveniles to adults in the population. Other characteristics that can be influenced by population interactions include the proportion of individuals within a population that have a specific trait or the average height or weight of the members of that population. It is therefore important for biologists to understand how different types of interactions can result in a change in the characteristics of a population.

One type of interaction that can result in a change in the characteristics of a population is predation. Predation often has a strong influence on the characteristics of a prey population. For example, a population of herbivores that lives in an area with a lot of predators will often have different characteristics than a population of herbivores that lives in an area with few or no predators. The hunting strategy used by the predator will also have an influence on the characteristics of a prey population. For example, a herbivore population that is eaten by a predator that chases its prey and a herbivore population that is eaten by a predator that hunts by sitting and waiting for its prey will often have different characteristics. Biologists often study how the characteristics of a specific prey population change in response to a specific type of predation, to understand how different types of interactions can result in a change in the characteristics of a population.

LAB 23

Your Task

Use a computer simulation called *Bug Hunt Speed* to explore how a population of a "bug" responds to the influence of two different types of predators. You will then develop an explanation for the changes you observe in the bug population. Your explanation must outline a mechanism that will cause the characteristics of a prey population to change in different ways in response to different types of predation.

The guiding question of this investigation is, **Why will the characteristics of a bug population change in different ways in response to different types of predation?**

Materials

You will use an online simulation called *Bug Hunt Speed* to conduct your investigation. You can access the simulation by going to the following website: *http://ccl.northwestern.edu/netlogo/models/BugHuntSpeeds*.

Safety Precautions

1. Use caution when working with electrical equipment. Keep away from water sources in that they can cause shorts, fires and shock hazards. Use only GFI protected circuits.
2. Wash hands with soap and water after completing this lab.
3. Follow all normal lab safety rules.

Getting Started

Bug Hunt Speed simulates a population of bugs that all belong to the same species. All the bugs in this population, however, are different, even though they belong to the same species. The bugs vary in terms of color, how fast they move, if they wiggle or not, and if they flee from predators when one is nearby.

In this simulation, you will act as the predator. You are able to eat the bugs (your prey) by clicking on them. You can act as a "hunting" predator by moving the mouse around to catch the bugs, or you can act as a "sit and wait" predator by keeping the mouse in one place and then catching the bugs that come to you. When a bug is eaten, it is replaced through reproduction by a bug in the simulated ecosystem. The new bug, therefore, will have the same characteristics as a bug that has not been eaten yet. Remember, all of the bugs in the ecosystem are from the same species.

The simulation also allows you to adjust the characteristics of the bugs. You can use the menus on the left of the screen to determine the color scheme for the bugs, the initial number of bugs in the habitat, if the bugs wiggle or not, and if they "flee" from a predator or not (see the figure on the opposite page).

Mechanisms of Evolution
Why Will the Characteristics of a Bug Population Change in Different Ways in Response to Different Types of Predation?

A screen shot from the *Bug Hunt Speed* simulation

To answer the guiding question, you must determine what type of data you will need to collect, how you will collect it, and how you will analyze it. To determine *what type of data you will need to collect*, think about the following questions:

- How will you determine if the characteristics of the bug population change over time?
- How will you test your explanation for the changes you observe in the population of bugs?
- What will serve as your dependent variable (e.g., color, speed, number of bugs caught)?
- What type of measurements or observations will you need to record during your investigation?

To determine *how you will collect your data*, think about the following questions:

- What will serve as a control condition (e.g., no predation)?
- What types of treatment conditions will you need to set up and how will you do it?
- How many trials will you need to conduct?
- How long will you need to run the simulation during each trial (e.g., for three minutes or until 60 bugs are caught)?

LAB 23

- How often will you collect data and when will you do it?
- How will you keep track of the data you collect and how will you organize the data?

To determine *how you will analyze your data,* think about the following questions:

- How will you determine if there is a difference between the different treatment conditions and the control condition?
- What type of calculations will you need to make?
- What type of graph could you create to help make sense of your data?

Investigation Proposal Required?   ☐ Yes   ☐ No

Connections to Crosscutting Concepts and to the Nature of Science and the Nature of Scientific Inquiry

As you work through your investigation, be sure to think about

- the importance of looking for patterns in nature,
- the importance of developing causal explanations for natural phenomena,
- the different types of methods that scientists use to answer questions, and
- the difference between observations and inferences.

Argumentation Session

Once your group has finished collecting and analyzing your data, prepare a whiteboard that you can use to share your initial argument. Your whiteboard should include all the information shown in the figure below.

Argument presentation on a whiteboard

The Guiding Question:	
Our Claim:	
Our Evidence:	Our Justification of the Evidence:

To share your argument with others, we will be using a round-robin format. This means that one member of your group will stay at your lab station to share your group's argument while the other members of your group go to the other lab stations one at a time to listen to and critique the arguments developed by your classmates.

The goal of the argumentation session is not to convince others that your argument is the best one; rather, the goal is to identify errors or instances of faulty reasoning in the arguments so these mistakes can be fixed. You will therefore need to evaluate the content of the claim, the quality of the evidence used to support the claim, and the strength of the justification of the evidence included in each argument that you see. In order to critique an argument, you will need more information than what is included on the whiteboard. You might, therefore, need to ask the presenter one or more follow-up questions, such as:

- How did you use the simulation to collect your data?
- What did you do to analyze your data? Why did you decide to do it that way? Did you check your calculations?
- Is that the only way to interpret the results of your analysis? How do you know that your interpretation of your analysis is appropriate?
- Why did your group decide to present your evidence in that manner?
- What other claims did your group discuss before you decided on that one? Why did your group abandon those alternative ideas?
- How confident are you that your claim is valid? What could you do to increase your confidence?

Once the argumentation session is complete, you will have a chance to meet with your group and revise your original argument. Your group might need to gather more data or design a way to test one or more alternative claims as part of this process. Remember, your goal at this stage of the investigation is to develop the most valid or acceptable answer to the research question!

Report

Once you have completed your research, you will need to prepare an investigation report that consists of three sections that provide answers to the following questions:

1. What question were you trying to answer and why?
2. What did you do during your investigation and why did you conduct your investigation in this way?
3. What is your argument?

Your report should answer these questions in two pages or less. This report must be typed, and any diagrams, figures, or tables should be embedded into the document. Be sure to write in a persuasive style; you are trying to convince others that your claim is acceptable or valid!

LAB 23

Lab 23. Mechanisms of Evolution: Why Will the Characteristics of a Bug Population Change in Different Ways in Response to Different Types of Predation?

Checkout Questions

1. There are two varieties of moles, brown and white, living on an island. They are a source of food for the owls. The island recently had a volcanic eruption and is now covered with dark ash, dark volcanic rock, and some soil. What will be the effect on the mole population over time?

2. Scientists often make different inferences based on the same observations.

   a. I agree with this statement.
   b. I disagree with this statement.

   Explain your answer, using information from your investigation about the mechanisms of evolution.

3. All scientists use the same method to test their ideas.

   a. I agree with this statement.

b. I disagree with this statement.

Explain your answer, using examples from your investigation about the mechanisms of evolution.

4. Scientists often attempt to identify patterns in nature. Explain why the identification of patterns is useful in science, using an example from your investigation about the mechanisms of evolution.

5. An important goal in science is to develop explanations for natural phenomena. Explain why the development of explanations is so important in science, using an example from your investigation about the mechanisms of evolution.

LAB 24

Lab 24. Descent With Modification: Does Mammalian Brain Structure Support or Refute the Theory of Descent With Modification?

Lab Handout

Introduction

One of Darwin's most revolutionary ideas was that all living things are related. According to Darwin, all life on Earth today is actually related because all life on Earth shares a common ancestor. This ancestor, he argued, lived on Earth sometime in the distant past but is now extinct. As a result of this common ancestor, all organisms now share some common features. An example of how organisms share common features can be seen in the figure below. The figure illustrates how the limbs of mammals have a similar internal structure. These limbs are examples of homologous structures because they have a similar internal structure but serve different functions (such as walking, swimming, or grasping).

Bones found in the limbs of seven different mammals

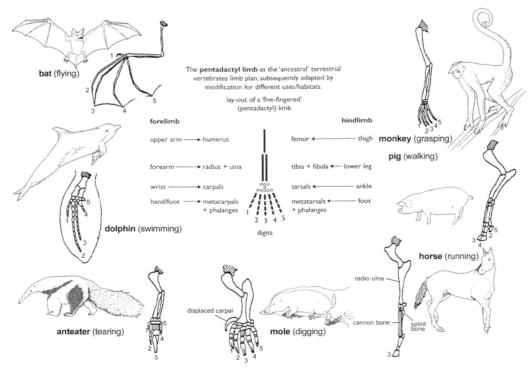

Descent With Modification
Does Mammalian Brain Structure Support or Refute the Theory of Descent With Modification?

To explain the similar internal bone structure in these mammals, Darwin said that they must be descendants of the same ancestor that had a limb that consisted of the same set of bones. He reasoned that the difference in the shape of the bones was a result of gradual modifications that made the organisms better adapted to survive in a particular environment. He called this idea descent with modification. He argued that environmental factors, over time, could slowly select for or against subtle variations in the basic shape of the bones in the limbs of organisms but could not completely change the basic body plan. This selection process would gradually result in whale fins and bat wings that had fingers similar to the fingers of a lizard, a frog, or a monkey. These variations would give their owners an advantage in a particular environment, such as the ocean in the case of the whale or the air in the case of the bat.

According to this theory, all species will share some physical features because all species share a common ancestor. This theory, however, also predicts that two species that have diverged from one another relatively recently in time will share more features than species that diverged from one another earlier. This prediction is based on the assumption that the longer it has been since two species separated from the same ancestral species, the more time there will be for differences to accumulate in each independent line.

The theory of descent with modification certainly seems reasonable. However, like all theories in science, the principles of descent with modification must be tested in many different ways before it can be considered valid or acceptable by the scientific community. In this investigation, you will determine if mammalian brain structure is consistent with the principles of descent with modification.

Your Task

Collect data about the brain structure of at least 10 different mammals. Then use the data you collect to test the theory of descent with modification.

The guiding question of this investigation is, **Does mammalian brain structure support or refute the theory of descent with modification?**

Materials

You will use an online database called Comparative Mammalian Brain Collections to conduct your investigation. You can access the database by going to the following website: *http://brainmuseum.org*.

Safety Precautions

1. Use caution when working with electrical equipment. Keep away from water sources in that they can cause shorts, fires, and shock hazards. Use only GFI-protected circuits.

LAB 24

2. Wash hands with soap and water after completing this lab.
3. Follow all normal lab safety rules.

Getting Started

To answer the guiding question, you will need to compare and contrast the brains from a sample of at least 10 different mammals. You MUST include the polar bear, the domestic dog, the domestic guinea pig, and the gorilla in your sample. To compare and contrast these brains, you will need to access the online database Comparative Mammalian Brain Collections, which is sponsored by the University of Wisconsin, Michigan State University, and the National Museum of Health and Medicine.

Once you have accessed the database, you must determine what type of data you will need to collect, how you will collect it, and how you will analyze it. To determine *what type of data you will need to collect,* think about the following questions:

- What would you expect mammalian brains to look like if the theory of descent with modification is valid? What would you expect mammalian brains to look like if the theory of descent with modification is not valid? (Hint: Think about what the brains of mammals would look like if mammals did not share a common ancestor.)
- How will you be able to identify the major structures of the mammalian brain, and how will you determine the function of each structure? (Hint: The database includes information about the structure and function of mammalian brains. You can access this information at *http://brainmuseum.org/functions/index.html*.)
- Which characteristics of the mammalian brain will you examine?
- How many different characteristics of the mammalian brain will you need to examine? (Hint: You should examine at least four different characteristics of each brain.)

To determine *how you will collect the data you need,* think about the following questions:

- How will you quantify differences and similarities in brain characteristics? (Hint: If you decide to examine the texture of the brain, you could look at the presence or absence of folds on the surface of the brain. If you decide to examine the shape of the brain, you could calculate the height-to-length ratio. If you decide to examine the size of different structures found in the brain, you could calculate a ratio between the length of a particular structure and the overall length of the brain.)
- How will you make sure that your data are of high quality (i.e., what will you do to help reduce measurement error)?
- What will you do with the data you collect?

To determine *how you will analyze your data,* think about the following questions:

Descent With Modification
Does Mammalian Brain Structure Support or Refute the Theory of Descent With Modification?

- How will you compare and contrast the various brains?
- What type of graph or table could you create to help make sense of your data?

Investigation Proposal Required?   ☐ Yes   ☐ No

Connections to Crosscutting Concepts and to the Nature of Science and the Nature of Scientific Inquiry

As you work through your investigation, be sure to think about

- the importance of looking at proportional relationships in science,
- the relationship between structure and function in nature,
- how science as a body of knowledge develops over time, and
- the different types of methods that scientists use to answer questions.

Argumentation Session

Once your group has finished collecting and analyzing your data, prepare a whiteboard that you can use to share your initial argument. Your whiteboard should include all the information shown in the figure below.

Argument presentation on a whiteboard

The Guiding Question:	
Our Claim:	
Our Evidence:	Our Justification of the Evidence:

To share your argument with others, we will be using a round-robin format. This means that one member of your group will stay at your lab station to share your group's argument while the other members of your group go to the other lab stations one at a time to listen to and critique the arguments developed by your classmates.

The goal of the argumentation session is not to convince others that your argument is the best one; rather, the goal is to identify errors or instances of faulty reasoning in the arguments so these mistakes can be fixed. You will therefore need to evaluate the content of the claim, the quality of the evidence used to support the claim, and the strength of the justification of the evidence included in each argument that you see. In order to critique an argument, you will need more information than what is included on the whiteboard. You might, therefore, need to ask the presenter one or more follow-up questions, such as:

- How did you use the database to collect your data? Why did you decide to do it that way? Why did you focus on those features of the brain?
- What did you do to make sure the data you collected are reliable? What did you do to decrease measurement error?

LAB 24

- What did you do to analyze your data? Why did you decide to do it that way? Did you check your calculations?
- Is that the only way to interpret the results of your analysis? How do you know that your interpretation of your analysis is appropriate?
- Why did your group decide to present your evidence in that manner?
- What other claims did your group discuss before you decided on that one? Why did your group abandon those alternative ideas?
- How confident are you that your claim is valid? What could you do to increase your confidence?

Once the argumentation session is complete, you will have a chance to meet with your group and revise your original argument. Your group might need to gather more data or design a way to test one or more alternative claims as part of this process. Remember, your goal at this stage of the investigation is to develop the most valid or acceptable answer to the research question!

Report

Once you have completed your research, you will need to prepare an investigation report that consists of three sections that provide answers to the following questions:

1. What question were you trying to answer and why?
2. What did you do during your investigation and why did you conduct your investigation in this way?
3. What is your argument?

Your report should answer these questions in two pages or less. This report must be typed, and any diagrams, figures, or tables should be embedded into the document. Be sure to write in a persuasive style; you are trying to convince others that your claim is acceptable or valid!

Lab 24. Descent With Modification: Does Mammalian Brain Structure Support or Refute the Theory of Descent With Modification?

Checkout Questions

1. What are the basic principles of the theory of descent with modification?

2. Use the theory of descent with modification to explain why the number of neck vertebrae is the same in all mammals. The neck of a giraffe, for example, is made of seven long bones, and the neck of the whale is made of seven short bones.

LAB 24

3. All scientific investigations must follow the same step-by-step method to be considered scientific.

    a. I agree with this statement.
    b. I disagree with this statement.

    Explain your answer, using examples from your investigation about descent with modification.

4. Scientific knowledge does not change.

    a. I agree with this statement.
    b. I disagree with this statement.

    Explain your answer, using information from your investigation about descent with modification.

Descent With Modification
Does Mammalian Brain Structure Support or Refute the Theory of Descent With Modification?

5. Scientists often need to look for proportional relationships when they analyze data or make comparisons. Explain why proportional relationships are useful in science, using an example from your investigation about descent with modification.

6. Structure and function are related in nature. Explain why, using an example from your investigation about descent with modification.

LAB 25

Lab 25. Mechanisms of Speciation: Why Does Geographic Isolation Lead to the Formation of a New Species?

Lab Handout

Introduction

There have been a number of models that have been proposed to explain the process of speciation (i.e., the formation of a new species). One of these models, called allopatric speciation, suggests that new species can arise when a population is divided into two or more subpopulations by some type of geographic barrier (such as a mountain range or an ocean). The two isolated subpopulations then diverge into different species over time (see the figure to the right). This explanation of the model, however, is incomplete because it only suggests that two or more subpopulations can become separate species when a geographic barrier separates them; it does not explain what causes the two subpopulations to diverge into different species over time.

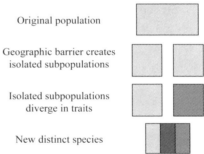

The model of allopatric speciation

Original population

Geographic barrier creates isolated subpopulations

Isolated subpopulations diverge in traits

New distinct species

Your Task

Use a computer simulation called *Bug Hunt Camouflage* to develop an explanation for how geographic isolation could lead to the formation of a new species. Your goal is to identify an underlying mechanism that can cause a physical characteristic (such as body color) found in two different populations of the same species to diverge when they live in different environments (seashore, glacier, or poppy field). This mechanism must be able to change a characteristic in the two populations enough so the individual members of one population will no longer interbreed with members of the other population.

The guiding question of this investigation is, **Why does geographic isolation lead to the formation of a new species?**

Materials

You will use the following materials during your investigation:

- An online simulation called *Bug Hunt Camouflage*, which can be accessed at *http://ccl.northwestern.edu/netlogo/models/BugHuntCamouflage*.
- Natural Selection and Species Concept Fact Sheet

Mechanisms of Speciation
Why Does Geographic Isolation Lead to the Formation of a New Species?

Safety Precautions

1. Use caution when working with electrical equipment. Keep away from water sources in that they can cause shorts, fires and shock hazards. Use only GFI protected circuits.

2. Wash hands with soap and water after completing this lab. Follow all normal lab safety rules.

Getting Started

Bug Hunt Camouflage simulates a population of bugs that all belong to the same species. However, the bugs in this population are not all the same color. You will be acting as a predator by using the mouse to eat the bugs (your prey) by clicking on them. When a bug is eaten, it is replaced through reproduction by a bug in the simulated environment. The new bug will appear near the parent bug and will start out as a small dot and then grow after a few seconds. The new bug will usually have the same characteristics as one of the bugs that have not been eaten yet. Some offspring, however, will have a slightly different coloration than the parent it came from because the simulation will allow the pigment genes to drift (to simulate the effect of random mutation). You can also create an entire new generation of bugs at any time by clicking the "Make a Generation" button. When you click this button, all of the bugs in the environment produce one offspring.

The simulation also allows you to adjust the following factors (see the figure on p. 204):

- The number of bugs in the environment (0 bugs to 100 bugs)
- The size of bugs in the environment (0.5 to 5)
- The nature of the environment (seashore, glacier, or poppy field)
- How much the pigment gene can drift from the parent value when a new bug is produced (0 to 98)

The simulation also allows you to keep track of a number of characteristics of the bugs and the environment over time:

- How many bugs you have caught (total-caught)
- How many bugs are in the world (bugs in world)
- Your progress and performance as a predator (bugs caught vs. time)
- How the average values for the hue, saturation, and brightness of the bugs change over time (average HSB values)
- The distribution of hues in the starting population (initial hues)
- The distribution of hues in the current population (current hues)

LAB 25

A screen shot from the *Bug Hunt Camouflage* simulation

- The distribution of saturations (of colors) in the current population (current saturation) and starting populations (initial saturation). Low values represent "grayish" colorations and high values represent "vivid" colorations.
- The distribution of brightness (of colors) in the current (current brightness) and starting (initial brightness) populations. Low values represent "dark" colorations and high values represent "light" colorations.
- How the average values of the genotype of the population change over time (vector difference in average genotype). The plot shows the vector difference between the average value of red gene frequency, green gene frequency, and blue gene frequency for the current population compared with the initial population.

To answer the guiding question using this computer simulation, you must determine what type of data you will need to collect, how you will collect it, and how you will analyze it. To determine *what type of data you will need to collect*, think about the following questions:

- How will you determine if the characteristics of a bug population change over time?
- What will serve as your dependent variable (e.g., number of bugs caught, hues, brightness, saturation)?
- What type of measurements or observations will you need to record during your investigation?

Mechanisms of Speciation
Why Does Geographic Isolation Lead to the Formation of a New Species?

To determine *how you will collect your data*, think about the following questions:

- What will serve as a control condition (glacier, poppy field, seashore)?
- What types of treatment conditions will you need to set up and how will you do it?
- What factors will you need to keep constant during each simulation?
- What should the characteristics of the initial population of bugs be at the beginning of each simulation?
- How long will you need to run the simulation (e.g., for three minutes or until 60 bugs are caught)?
- How many trials will you need to conduct for each treatment?
- How often will you collect data and when will you do it?
- How will you keep track of the data you collect and how will you organize the data?

To determine *how you will analyze your data*, think about the following questions:

- How will you determine if the environment affected the characteristics of a bug population?
- What type of calculations will you need to make?
- What type of graph could you create to help make sense of your data?

Investigation Proposal Required?   ☐ Yes   ☐ No

Connections to Crosscutting Concepts and to the Nature of Science and the Nature of Scientific Inquiry

As you work through your investigation, be sure to think about

- the importance of identifying causal relationships in science,
- the importance of looking at proportional relationships in science,
- how scientists use models,
- how theories are developed rather than discovered, and
- the different types of methods that scientists use to test ideas.

Argumentation Session

Once your group has finished collecting and analyzing your data, prepare a whiteboard that you can use to share your initial argument. Your whiteboard should include all the information shown in the figure on p. 206.

To share your argument with others, we will be using a round-robin format. This means that one member of your group will stay at your lab station to share your group's argument while the other members of your group go to the other lab stations one at a time

LAB 25

Argument presentation on a whiteboard

The Guiding Question:	
Our Claim:	
Our Evidence:	Our Justification of the Evidence:

to listen to and critique the arguments developed by your classmates.

The goal of the argumentation session is not to convince others that your argument is the best one; rather, the goal is to identify errors or instances of faulty reasoning in the arguments so these mistakes can be fixed. You will therefore need to evaluate the content of the claim, the quality of the evidence used to support the claim, and the strength of the justification of the evidence included in each argument that you see. In order to critique an argument, you will need more information than what is included on the whiteboard. You might, therefore, need to ask the presenter one or more follow-up questions, such as:

- How did you use the simulation to collect your data?
- What did you do to analyze your data? Why did you decide to do it that way? Did you check your calculations?
- Is that the only way to interpret the results of your analysis? How do you know that your interpretation of your analysis is appropriate?
- Why did your group decide to present your evidence in that manner?
- What other claims did your group discuss before you decided on that one? Why did your group abandon those alternative ideas?
- How confident are you that your claim is valid? What could you do to increase your confidence?

Once the argumentation session is complete, you will have a chance to meet with your group and revise your original argument. Your group might need to gather more data or design a way to test one or more alternative claims as part of this process. Remember, your goal at this stage of the investigation is to develop the most valid or acceptable answer to the research question!

Report

Once you have completed your research, you will need to prepare an investigation report that consists of three sections that provide answers to the following questions:

1. What question were you trying to answer and why?
2. What did you do during your investigation and why did you conduct your investigation in this way?
3. What is your argument?

Mechanisms of Speciation
Why Does Geographic Isolation Lead to the Formation of a New Species?

Your report should answer these questions in two pages or less. This report must be typed, and any diagrams, figures, or tables should be embedded into the document. Be sure to write in a persuasive style; you are trying to convince others that your claim is acceptable or valid!

LAB 25

Natural Selection and Species Concept Fact Sheet

The Theory of Natural Selection

The fossil record provides convincing evidence that species evolve. In other words, the number of species found on Earth and the characteristics of these species have changed over time. However, these observations tell us little about the natural processes that drive evolution. A number of different explanations have been offered by scientists in an effort to explain why (or if) evolution occurs. One of these explanations is called natural selection. The basic tenets of natural selection are as follows (Lawson 1995):

- Only a fraction of the individuals that make up a population survive long enough to reproduce.
- The individuals in a population are not all the same. Individuals have traits that make them unique.
- Much, but not all, of this variation in traits is inheritable and can therefore be passed down from parent to offspring.
- The environment, including both abiotic (e.g., temperature, amount of water available) and biotic (e.g., amount of food, presence of predators) factors, determines which traits are favorable or unfavorable, because some traits increase an individual's chance of survival and others do not.
- Individuals with favorable traits tend to produce more offspring than those with unfavorable traits. Therefore, over time, favorable traits become more common within a population found in a particular environment (and unfavorable traits become less common).

How Biologists Define a Species

A species can be defined as "a population or group of populations whose members have the potential to interbreed with one another in nature to produce viable, fertile offspring, but who cannot produce viable, fertile offspring with members of other species" (Campbell and Reece 2002, p. 465). This definition is known as the biological species concept. A group of individuals can therefore be classified as a species when there are one or more factors that will prevent them from interbreeding with individuals from another group.

In nature, however, the biological species concept does not always work well. A bacterium, for example, reproduces by copying its genetic material and then splitting (this is called binary fission). Therefore, defining a species as a group of interbreeding individuals only works with organisms that do not use an asexual form of reproduction. Most plants (and some animals) that use sexual reproduction can also self-fertilize, which makes it difficult to determine the boundaries of a species. Biologists are also unable to check for the ability to interbreed in extinct forms of organisms found in the fossil record.

Therefore, many other "species concepts" have been proposed by scientists, such as the ecological species concept (which means a species is defined by its ecological niche or its role in a biological community), the morphological species concept (which means a species is defined using a unique set of shared structural features), and the genealogical species concept (which means a species is a set of organisms with a unique genetic history). The species concept that a scientist chooses to use will often reflect his or her research focus. All scientists, however, are expected to decide on a species concept, provide a rationale for doing so, and then use it consistently.

References

Benton, M. J. 1995. Diversification and extinction in the history of life. *Science* 268 (5207): 52–58.

Campbell, N., and J. Reece. 2002. *Biology*. 6th ed. San Francisco: Benjamin Cummings.

Lawson, A. 1995. *Science teaching and the development of thinking*. Belmont, CA: Wadsworth.

LAB 25

Lab 25. Mechanisms of Speciation: Why Does Geographic Isolation Lead to the Formation of a New Species?

Checkout Questions

1. How can geographic isolation result in the formation of a new species?

2. Experiments are a necessary part of the scientific process. Without an experiment, a study is not rigorous or scientific.

    a. I agree with this statement.
    b. I disagree with this statement.

    Explain your answer, using information from your investigation about the mechanisms of speciation.

3. Scientific theories exist in the natural world and are uncovered through scientific investigations (i.e., scientists discover theories).

    a. I agree with this statement.
    b. I disagree with this statement.

Mechanisms of Speciation
Why Does Geographic Isolation Lead to the Formation of a New Species?

Explain your answer, using examples from your investigation about the mechanisms of speciation.

4. Scientists often need to look for proportional relationships when they analyze data or make comparisons. Explain why this is important for scientists to do, using an example from your investigation about the mechanisms of speciation.

5. Structure and function are related in nature. How did this principle play a role in your investigation about the mechanisms of speciation?

LAB 26

Lab 26. Human Evolution: How Are Humans Related to Other Members of the Family Hominidae?

Lab Handout

Introduction

The central idea of biological evolution is that all life on Earth shares a common ancestor. All organisms found on Earth are therefore related, and their unique features are the result of the process of descent with modification. Scientists can determine which species are most closely related by studying the unique inheritable characteristics these species share and other historical information.

Biologists use phylogenetic trees to represent evolutionary relationships between organisms. A phylogenetic tree is a branching diagram that shows the inferred evolutionary relationships among various biological entities based on similarities and differences in their physical and/or genetic characteristics. The root of the phylogenetic tree represents the common ancestor, and the tips of the branches represent all the descendants (see the left figure below). As you move from the tips to the root of the tree, you are moving backward in time. The branches on the tree represent speciation events (see the right figure below). When a speciation event occurs, a single species (ancestral lineage) gives rise to two or more new species (daughter lineages).

The root of a phylogenetic tree represents the common ancestor (left), and the branches represent speciation events (right).

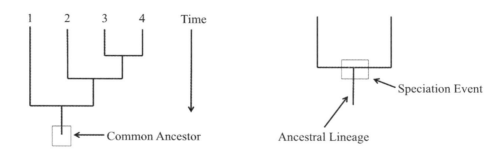

Phylogenetic trees are used to show common ancestry. Each species has a part of its history that is unique to it alone and parts that are shared with other species (see the top left figure on the opposite page). Similarly, each species has ancestors that are unique to that species and ancestors that are shared with other species (see the top right figure on the opposite page).

Human Evolution
How Are Humans Related to Other Members of the Family Hominidae?

Each species has both unique and shared histories (left) and unique and shared ancestors (left).

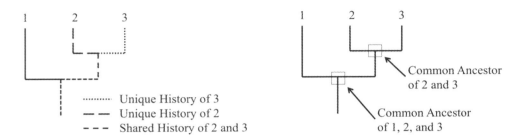

A clade is a grouping that includes a common ancestor and all the descendants (living and extinct) of that ancestor. Clades are nested within one another—scientists call this a nested hierarchy. A clade may include thousands of species or just a few. Some examples of clades at different levels are marked in the phylogenetic tree shown in the figure on the left below. Notice how clades are nested within larger clades. Biologists often represent time on phylogenies by drawing the branch lengths in proportion to the amount of time that has passed since that lineage arose. The figure on the right below provides an example of how a phylogenetic tree can be used to illustrate when different lineages arose or went extinct in the history of life on Earth (MYA in this figure stands for "million years ago," so [for example] "100" means "100 million years ago").

A clade (left) includes a common ancestor and all of its descendants. Phylogenetic trees can show when different species arose or went extinct.

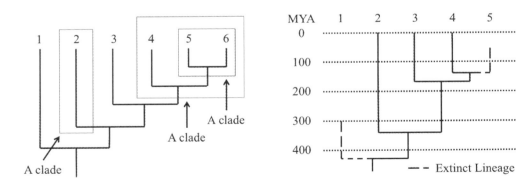

Humans and other members of the family Hominidae are notable among the rest of the primates for their bipedal locomotion, slow rate of maturation, and large brain size. Our current understanding of the evolutionary history of Hominidae is derived largely from the findings of paleontology and anthropology. Thousands of fossils of human ancestors and extinct relatives have been unearthed. Each fossil, whether it is a complete skeleton

LAB 26

or a single tooth, contributes significantly to our understanding of the origins of humans. In this investigation, you will have an opportunity to explore the evolutionary history of Hominidae.

Your Task

Develop a phylogenetic tree for the living and extinct members of the family Hominidae and then use this information to draw inferences about the evolutionary history of humans.

The guiding question for this investigation is, **How are humans related to other members of the family Hominidae?**

Materials

You will be supplied with a set of seven skulls. The table below provides some information about them.

Information about the Hominidae skulls

Name	Oldest specimen	Youngest specimen	Where found
Paranthropus boisei	2.3 MYA	1.2 MYA	Ethiopia, Tanzania, Kenya
Australopithecus afarensis	3.6 MYA	2.9 MYA	Ethiopia
Pan troglodytes (chimpanzee)	4.9 MYA	Today	Africa
Gorilla gorilla (gorilla)	5.9 MYA	Today	Africa
Homo erectus	2 MYA	400 TYA	Africa, Asia, and Europe.
Homo neanderthalensis	250 TYA	45 TYA	Europe and the Middle East
Homo sapiens	200 TYA	Today	Worldwide

Key: MYA = million years ago; TYA = thousand years ago.

Safety Precautions

1. Use caution when working with electrical equipment. Keep away from water sources in that they can cause shorts, fires and shock hazards. Use only GFI protected circuits.

2. Use caution when handling skulls—sharp edges can cut skin.

3. Wash hands with soap and water after completing this lab.

4. Follow all normal lab safety rules.

Human Evolution
How Are Humans Related to Other Members of the Family Hominidae?

Getting Started

Your first step is to carefully examine the seven Hominidae skulls to identify the similarities and differences between them. The figure below lists 13 features that scientists use to describe a hominid skull. You should collect data about all 13 of these features for each skull.

Features of Hominidae skulls

Aspect of the skull	Location
Braincase and face: • Presence or absence of a supraorbital browridge • Continuous or divided supraorbital browridge • Size of the braincase • Presence or absence of a sagittal crest • Flat or protruding mastoid process • Raised or flat nasal bones • Maximum height of the nasal opening (in millimeters) • Length of the mandible (in millimeters) • Slope of the face (in degrees)	Labeled diagram of skull showing: Supraorbital Browridge, Sagittal Crest, Nasal Bone, Nasal Opening, Slope of Face, Mastoid Process, Length of Mandible
Dentition: • Combined width or breadth of the four incisors (in millimeters) • Canine tooth that protrudes above the chewing surfaces of the other teeth • Presence or absence of a canine diastema (a space between the canine tooth and the incisors) • Combined length of the two premolars and the three molars (in millimeters)	Labeled diagram of teeth showing: 3 Molars, 2 Premolars, Canine Tooth, Incisors

Once you have collected your data, you will need to analyze it. One way to accomplish this goal is to create a phylogenetic tree. Biologists begin the process of constructing a phylogenetic tree by collecting data about heritable traits that can be compared across species (such as the 13 skull features listed in the figure above). Biologists then determine which species have a shared derived character and which do not. A shared derived character is one that two species have in common and that has appeared in the lineage leading up to a clade. As a result, the character sets members of a clade apart from other individuals not in the clade. It is important to note, however, that a shared derived character does not need to be a new feature or a feature that has increased in size; a shared derived character can also be a feature that has gotten smaller or has disappeared completely. For example, the absence of a sagittal crest might be a shared derived character in hominids that sets the members of a clade apart from other clades. One way to keep track of the presence or absence of the shared derived characters for each hominid is to create a chart such

LAB 26

as the one shown in the "Features of the hominidae skull" figure.

Biologists then create the phylogenetic tree by grouping species into clades based on the number of shared derived characters they have. The figure to the right shows an example of a phylogenetic tree that is based on the number of derived characters shared by species 1–6 in Table 2. Notice that species 1 has the fewest number of shared derived characters (only A) and species 3–6 have the most (A, B, C, and D or A, B, C, and E). Species 3 and 4 share different characters (A, B, C, and D) than 5 and 6 (A, B, C, and E)—this means that species 3 and 4 share more of their evolutionary history than species 5 and 6 do, even though they have the same number of shared derived characters.

Presence of shared derived characters for each species

Species	Shared derived character					
	A	B	C	D	E	Total
1	Y	N	N	N	N	1
2	Y	Y	N	N	N	2
3	Y	Y	Y	Y	N	4
4	Y	Y	Y	Y	N	4
5	Y	Y	Y	N	Y	4
6	Y	Y	Y	N	Y	4

A phylogenetic tree

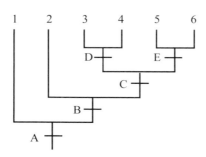

Once you have created your phylogenetic tree, you will use it as evidence to support your answer to the guiding question.

Investigation Proposal Required? ☐ Yes   ☐ No

Connections to Crosscutting Concepts and to the Nature of Science and the Nature of Scientific Inquiry

As you work through your investigation, be sure to think about

- the importance of identifying and explaining patterns in science,
- the relationship between structure and function in nature,
- the difference between data and evidence in science, and
- the different types of methods that scientists use to answer questions.

Human Evolution
How Are Humans Related to Other Members of the Family Hominidae?

Argumentation Session

Once your group has finished collecting and analyzing your data, prepare a whiteboard that you can use to share your initial argument. Your whiteboard should include all the information shown in the figure to the right.

To share your argument with others, we will be using a round-robin format. This means that one member of your group will stay at your lab station to share your group's argument while the other members of your group go to the other lab stations one at a time to listen to and critique the arguments developed by your classmates.

Argument presentation on a whiteboard

The Guiding Question:	
Our Claim:	
Our Evidence:	Our Justification of the Evidence:

The goal of the argumentation session is not to convince others that your argument is the best one; rather, the goal is to identify errors or instances of faulty reasoning in the arguments so these mistakes can be fixed. You will therefore need to evaluate the content of the claim, the quality of the evidence used to support the claim, and the strength of the justification of the evidence included in each argument that you see. In order to critique an argument, you will need more information than what is included on the whiteboard. You might, therefore, need to ask the presenter one or more follow-up questions, such as:

- How did you collect your data? Why did you decide to use that method? Why did you collect those data?
- What did you do to make sure the data you collected are reliable? What did you do to decrease measurement error?
- What did you do to analyze your data? Why did you decide to do it that way? Did you check your calculations?
- Is that the only way to interpret the results of your analysis? How do you know that your interpretation of your analysis is appropriate?
- Why did your group decide to present your evidence in that manner?
- What other claims did your group discuss before you decided on that one? Why did your group abandon those alternative ideas?
- How confident are you that your claim is valid? What could you do to increase your confidence?

Once the argumentation session is complete, you will have a chance to meet with your group and revise your original argument. Your group might need to gather more data or design a way to test one or more alternative claims as part of this process. Remember, your

LAB 26

goal at this stage of the investigation is to develop the most valid or acceptable answer to the research question!

Report

Once you have completed your research, you will need to prepare an investigation report that consists of three sections that provide answers to the following questions:

1. What question were you trying to answer and why?
2. What did you do during your investigation and why did you conduct your investigation in this way?
3. What is your argument?

Your report should answer these questions in two pages or less. This report must be typed, and any diagrams, figures, or tables should be embedded into the document. Be sure to write in a persuasive style; you are trying to convince others that your claim is acceptable or valid!

Lab 26. Human Evolution: How Are Humans Related to Other Members of the Family Hominidae?

Checkout Questions

1. Explain how humans and chimpanzees are related to members of the family Hominidae.

2. Evidence is data that have been analyzed and then interpreted by scientists.

    a. I agree with this statement.
    b. I disagree with this statement.

    Explain your answer, using examples from your investigation about human evolution.

LAB 26

3. All scientific investigations are experiments.

    a. I agree with this statement.
    b. I disagree with this statement.

    Explain your answer, using information from your investigation about evolution.

4. An important aspect of science is identifying and the explaining patterns in nature. Explain why this is important for scientists to do, using an example from your investigation about human evolution.

5. Scientists often need to think about how structure is related to function in nature when they analyze data or make comparisons. Explain why this is important for scientists to do, using an example from your investigation about human evolution.

Lab 27. Whale Evolution: How Are Whales Related to Other Mammals?

Lab Handout

Introduction

You have learned about the different categories of large molecules that play an important role in the bodies of organisms. One category of large molecule is called protein. A protein is made up of a chain of amino acids, and a specific gene determines the sequence of amino acids in the chain. Enzymes are examples of proteins that are very important to the function of an organism, and these proteins have very specific structures.

Scientists are often interested in the amino acid sequence of proteins for a number of reasons. Scientists, for example, often want to identify the amino acid sequence of a protein because the sequence determines the structure of a protein (see the figure to the right). Scientists can also use amino acid sequences to examine evolutionary relationships, because all life on Earth shares a common ancestor.

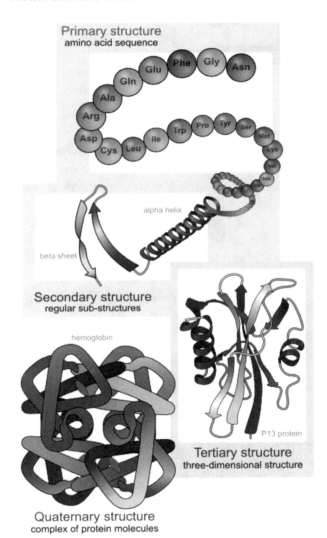

Protein structure levels

The central idea of biological evolution is that through a process of descent with modification, the common ancestor of all life on Earth gave rise to the biodiversity we see today (see the figure on p. 222). This idea is important in biology because it enables scientists to study the evolutionary history of life on Earth. The process of descent with modification, for example, suggests that two species that diverged from one another relatively recently in the history

LAB 27

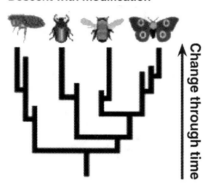

Descent with modification

of life on Earth will share more genetic similarities than two species that diverged from one another further back in time. Species that share many genetic similarities, as a result, are considered to be more closely related than two species that have many differences. Scientists, therefore, can compare an amino acid sequence for a specific protein to determine the evolutionary history of a group of organisms. In this investigation you will use an online database called UniProt that contains information about amino acid sequences to determine how whales are related to other mammals.

Your Task

Use the UniProt online database to examine the amino acid sequence for a protein found in all mammals. The protein you will examine is called hemoglobin subunit alpha (HBA); it enables red blood cells to transport oxygen. You will then choose several mammals that you think may be closely related to whales and use the UniProt database to create a phylogenetic tree that illustrates the evolutionary history of this group of mammals based on similarities and differences in the amino acid sequence of the HBA protein. You will then need to explain (1) how whales are related to other mammals and (2) how the phylogenetic tree you created supports your claim.

The guiding question of this investigation is, **How are whales related to other mammals?**

Materials

You will use the following materials during your investigation:

- An online database called UniProt, which can be accessed at *www.uniprot.org*
- Mammal Classification Fact Sheet

Safety Precautions

1. Use caution when working with electrical equipment. Keep away from water sources in that they can cause shorts, fires, and shock hazards. Use only GFI-protected circuits.//
2. Wash hands with soap and water after completing this lab.
3. Follow all normal lab safety rules.

Whale Evolution
How Are Whales Related to Other Mammals?

Getting Started

To answer the guiding question, you will need to use the UniProt database. This database is a collection of amino acid sequences that have been submitted by scientists from all over the world, and it is free for anyone to use. You will use this database to examine the HBA amino acid sequence found in whales and then compare it with the HBA amino acid sequence found in other mammals.

Once you access the UniProt database, follow these directions:

- In the "Query" box at the top of the page type in "HBA Whale" and click "Search."
  - Click on the first box to select "Sperm Whale."
  - Your selection should appear in a green toolbar at the bottom of the window.
- Go back up to the "Query" box at the top of the page and type in "HBA" and click "Search."
- Click on the boxes at the left of the page to make "checks" next to the organisms you want to compare with the sperm whale. Make sure you select animals with the gene HBA listed (not HBA1 or HBA2).
- Once you have made your selections, click on the "Align" button at the bottom of the page. It will take a few seconds to run, so be patient. You are running an application called ALIGN that compares the amino acid sequence of the organisms you selected. The application will provide you with the HBA amino acid sequence for all the organisms you selected, a phylogenetic tree based on differences in the sequence (which are shared derived characteristics), and a key that will tell you which entry belongs to which animal.

Once you know how to use the UniProt database to compare the amino acid sequences of different mammals, you need to think about what data you will need to collect and what you will do with the results of the analysis. To determine what type of data you need to collect, think about the following questions:

- Which mammals will you need to include in the analysis?
- How many mammals will you need to include?
- Will you choose mammals to represent different orders or to represent different families?

To determine what to do with the results of the analysis, think about the following questions:

- How will you keep track of the information you collect and how will you organize it?

LAB 27

- What would you expect to see in the amino acid sequence of a mammal that is closely related to the sperm whale? How about a mammal that is not very closely related to the sperm whale?
- Will you be able to use the results of your analysis as your claim, or are the results of your analysis part of the evidence you will use to support your claim?
- How can you share the results of your analysis with others?

Investigation Proposal Required?   ☐ Yes   ☐ No

Connections to Crosscutting Concepts and to the Nature of Science and the Nature of Scientific Inquiry

As you work through your investigation, be sure to think about

- the importance of identifying and explaining patterns in science,
- the relationship between structure and function in nature,
- the difference between data and evidence, and
- how science, as a body of knowledge, develops over time.

Argumentation Session

Once your group has finished collecting and analyzing your data, prepare a whiteboard that you can use to share your initial argument. Your whiteboard should include all the information shown in the figure below.

Argument presentation on a whiteboard

The Guiding Question:	
Our Claim:	
Our Evidence:	Our Justification of the Evidence:

To share your argument with others, we will be using a round-robin format. This means that one member of your group will stay at your lab station to share your group's argument while the other members of your group go to the other lab stations one at a time to listen to and critique the arguments developed by your classmates.

The goal of the argumentation session is not to convince others that your argument is the best one; rather, the goal is to identify errors or instances of faulty reasoning in the arguments so these mistakes can be fixed. You will therefore need to evaluate the content of the claim, the quality of the evidence used to support the claim, and the strength of the justification of the evidence included in each argument that you see. In order to critique an argument, you will need more information than what is included on the whiteboard. You might, therefore, need to ask the presenter one or more follow-up questions, such as:

- How did you use the database to collect your data? Why did you decide to focus on those mammals?
- What did you do to analyze your data? Why did you decide to do it that way? Did you check your calculations?
- Is that the only way to interpret the results of your analysis? How do you know that your interpretation of your analysis is appropriate?
- Why did your group decide to present your evidence in that manner?
- What other claims did your group discuss before you decided on that one? Why did your group abandon those alternative ideas?
- How confident are you that your claim is valid? What could you do to increase your confidence?

Once the argumentation session is complete, you will have a chance to meet with your group and revise your original argument. Your group might need to gather more data or design a way to test one or more alternative claims as part of this process. Remember, your goal at this stage of the investigation is to develop the most valid or acceptable answer to the research question!

Report

Once you have completed your research, you will need to prepare an investigation report that consists of three sections that provide answers to the following questions:

1. What question were you trying to answer and why?
2. What did you do during your investigation and why did you conduct your investigation in this way?
3. What is your argument?

Your report should answer these questions in two pages or less. This report must be typed, and any diagrams, figures, or tables should be embedded into the document. Be sure to write in a persuasive style; you are trying to convince others that your claim is acceptable or valid!

LAB 27

Mammal Classification Fact Sheet

Phylum	Class	Order*	Families†	Example
Chordata	Mammalia	Didelphimorphia	Didelphidae	Opossum
		Diprotodontia	Phascolarctidae	Koala
			Vombatidae	Wombat
			Macropodidae	Kangaroo
		Chiroptera	Pteropodidae	Fruit bat
			Emballonuridae	Sac-winged bat
			Vespertilionidae	Evening bat
		Primates	Lemuridae	Lemur
			Hominidae	Great ape
			Hylobatidae	Gibbon
		Carnivora	Felidae	Cat
			Canidae	Dog
			Ursidae	Bear
			Phocidae	Seal
			Odobenidae	Walrus
			Mustelidae	Weasel
		Cetacea	Balaenopteridae	Humpback whale
			Eschrichtiidae	Grey whale
			Physeteridae	Sperm whale
			Delphinidae	Dolphin
			Monodontidae	Beluga whale
			Phocoenidae	Porpoise
		Sirenia	Dugongidae	Dugong
			Trichechidae	Manatee
		Proboscidea	Elephantidae	Elephant
		Perissodactyla	Equidae	Horse
			Tapiridae	Tapir
			Rhinocerotidae	Rhinoceros
		Artiodactyla	Hippopotamidae	Hippopotamus
			Camelidae	Camel
			Giraffidae	Giraffe
			Cervidae	Deer
			Bovidae	Cow
		Rodentia	Castoridae	Beaver
			Caviidae	Guinea pig
			Cricetidae	Rat
			Sciuridae	Squirrel
		Lagomorpha	Leporidae	Rabbit

* Some orders have been omitted.

† Many of the families within each order have been omitted.

Lab 27. Whale Evolution: How Are Whales Related to Other Mammals?

Checkout Questions

1. How do the branches of a phylogenetic tree demonstrate the relatedness of species?

2. All evidence is data, but all data are not evidence.

   a. I agree with this statement.
   b. I disagree with this statement.

   Explain your answer, using examples from your investigation about whale evolution.

LAB 27

3. Scientific knowledge may be abandoned or modified in light of new evidence or due to the reconceptualization of prior evidence and knowledge.

    a. I agree with this statement.
    b. I disagree with this statement.

    Explain your answer, using information from your investigation about whale evolution.

4. An important aspect of science is identifying and explaining patterns in nature. Explain why this is important for scientists to do, using an example from your investigation about whale evolution.

5. Scientists often need to think about how structure is related to function in nature when they analyze data or make comparisons. Explain why this is important for scientists to do, using an example from your investigation about whale evolution.

IMAGE CREDITS

LAB 1
p. 13: Richard Wheeler, Wikimedia Commons, CC BY-SA 3.0. http://commons.wikimedia.org/wiki/File:Human_Erythrocytes_OsmoticPressure_PhaseContrast_Plain.svg

p. 15: Victor Sampson

LAB 2
p. 21: User:kaibara87, Wikimedia Commons, CC BY 2.0. http://commons.wikimedia.org/wiki/File%3AMouth_cells.jpg

LAB 3
p. 27: Plant: Luis Fernández Garcнa, Wikimedia Commons, CC BY-SA 2.5 ES. http://commons.wikimedia.org/wiki/File:Meristemo_apical_1.jpg; Animal: User:Staticd, Wikimedia Commons, CC BY-SA 3.0. http://commons.wikimedia.org/wiki/File:Onion_root_mitosis.jpg

LAB 4
p. 33: Photo by Pat Kenny (National Cancer Institute, National Institutes of Health, Wikimedia Commons, Public domain. http://commons.wikimedia.org/wiki/File:Normal_and_cancer_cells_%28labeled%29_illustration.jpg

LAB 5
p. 42: Victor Sampson

LAB 6
p.47 (left): Wikimedia Commons, Public domain. http://commons.wikimedia.org/wiki/File:Alpha-D-glucose-2D-skeletal-hexagon.png; (center) Wikimedia Commons, Public domain. http://commons.wikimedia.org/wiki/File:Lactose_Haworth.svg; (right) Wikimedia Commons, Public domain. http://commons.wikimedia.org/wiki/File:Amylopektin_Haworth.svg.

p. 48 (a): Wolfgang Schaefer, Wikimedia Commons, Public domain. http://commons.wikimedia.org/wiki/File:Fat_triglyceride_shorthand_formula.png; (b) Sten Andrй, Wikimedia Commons, Public domain. http://commons.wikimedia.org/wiki/File:Lysine_simple.png

p. 49: Victor Sampson

LAB 7
p. 55: H. McKenna, Wikimedia Commons, CC BY 2.5. http://commons.wikimedia.org/wiki/File:Leaf_anatomy.svg

p. 56: Victor Sampson

LAB 8
p. 61: Magnus Manske, Wikimedia Commons, CC BY-SA 3.0, GFDL 1.2. http://upload.wikimedia.org/wikipedia/commons/5/56/Enzyme_activation_energy.png

p. 62: User:Aejahnke, Wikimedia Commons, CC BY-SA 3.0. http://upload.wikimedia.org/wikipedia/commons/f/fc/Enzyme_mechanism_1.jpg

LAB 9
p. 71: National Archives of Australia, Wikimedia Commons, Public domain. http://commons.wikimedia.org/wiki/File:Rabbits_MyxomatosisTrial_WardangIsland_1938.jpg

p. 73: http://ccl.northwestern.edu/netlogo/models/RabbitsGrassWeeds

LAB 10
p. 79 (top): Doug Smith (National Park Service), Wikimedia Commons, Public domain. http://commons.wikimedia.org/wiki/File:Canis_lupus_pack_surrounding_Bison.jpg; (bottom): Victor Sampson

p. 81: http://ccl.northwestern.edu/netlogo/models/WolfSheepPredation via Victor Sampson

LAB 11
p. 88: Victor Sampson

p. 89: www.learner.org/courses/envsci/interactives/ecology/ecology1.html

LAB 12
p. 95: Terry Goss, Wikimedia Commons, CC BY-SA 3.0, GFDL. http://commons.wikimedia.org/wiki/File:White_shark.jpg

p. 97: www.ocearch.org via Victor Sampson

Image Credits

LAB 13
p. 103: Victor Sampson

LAB 14
p. 111: Victor Sampson

LAKE GRACE INFORMATION PACKET

p. 115: Victor Sampson

p. 118:

Largemouth bass: User:Solrman, Wikimedia Commons, CC0. http://commons.wikimedia.org,wiki,File%3ALargemouth-bass.jpg

white bass: Wikimedia Commons , Public domain , http://commons.wikimedia.org/wiki/File%3AWhite_Bass.jpg

Bluegill: National Oceanic and Atmospheric Administration, Wikimedia Commons , Public domain. http://commons.wikimedia.org/wiki/File%3ABlue_gill.jpg

Daphnia: Photo by Paul Hebert. Gewin, V. 2005. PLoS Biology 3 (6): e219, Wikimedia Commons, CC BY 2.5. http://commons.wikimedia.org/wiki/File%3ADaphnia_pulex.png

p. 119:

Gammarus amphipod: Michal Manas, Wikimedia Commons, CC BY 2.5. http://commons.wikimedia.org/wiki/File%3AGammarus_roeselii.jpg

algae: Vera Buhl, Wikimedia Commons, CC BY-SA 3.0, GFDL. http://commons.wikimedia.org/wiki/File%3A2009-08-19_(47)_Lake%2C_See.jpg

pickerelweed: Russ Pollanen / Wikimedia Commons / CC BY 2.5 / http://commons.wikimedia.org/wiki/File:Pickler-2-s.jpg

hydrilla: United States Geological Survey, Wikimedia Commons, Public domain. http://en.wikipedia.org/wiki/File:Hydrilla_USGS.jpg

water hyacinth: Ted Center (USDA), Wikimedia Commons, Public domain. http://commons.wikimedia.org/wiki/File%3AWater_hyacinth.jpg.

p. 121: Victor Sampson

LAB 15
p. 125: Mahesh Iyer, Wikimedia Commons, CC BY-SA 3.0, GFDL. http://commons.wikimedia.org/wiki/File:Eurasian_Collared_Dove.svg

p. 126: Victor Sampson

LAB 16
p. 136: www.fastplants.org/resources/digital_library//index.php?P=FullImage&ResourceId=35&FieldName=Screenshot via Victor Sampson

LAB 17
p. 144: Thomas Geier (Fachgebiet Botanik der Forschungsanstalt Geisenheim), Wikimedia Commons, CC BY-SA 3.0. http://commons.wikimedia.org/wiki/File%3AAllium-Mitose06-DM100x_BL28.jpg

LAB 18
DNA FACT SHEET

p. 153: User:cdang, Wikimedia Commons, CC BY-SA 3.0, GFDL. http://commons.wikimedia.org/wiki/File%3ACliche_de_laue_principe.svg; National Institutes of Health, Wikimedia Commons, Public domain. http://commons.wikimedia.org/wiki/File%3AXray_DNA.gif

LAB 19
p. 157: National Human Genome Research Institute, Wikimedia Commons, Public domain. http://commons.wikimedia.org/wiki/File:DNA_human_male_chromosomes.gif

p. 159: Victor Sampson

LAB 20
p. 164: Modified by Victor Sampson from User:InvictaHOG, Wikimedia Commons, Public domain. http://en.wikipedia.org/wiki/File:ABO_blood_type.svg

p. 166: Victor Sampson

LAB 21
p. 172: User:Madboy74, Wikimedia Commons, CC0 1.0. http://commons.wikimedia.org/wiki/File:Biology_Illustration_Animals_Insects_Drosophila_melanogaster.svg

Image Credits

LAB 22
p. 181: Peter Halasz, Wikimedia Commons, Public domain. *http://commons.wikimedia.org/wiki/File:Biological_classification_L_Pengo_vflip.svg*

LAB 23
p. 189: *http://ccl.northwestern.edu/netlogo/models/BugHuntSpeeds* via Victor Sampson

LAB 24
p. 194: Jerry Crimson Mann, Wikimedia Commons, CC BY-SA 3.0, GFDL 1.2. *http://commons.wikimedia.org/wiki/File:Evolution_pl.png.*

LAB 25
p. 202: Victor Sampson

p. 204: *http://ccl.northwestern.edu/netlogo/models/BugHuntCamouflage* via Victor Sampson

LAB 26
pp. 212 and 213: Victor Sampson

p. 215: Modified by Victor Sampson from Patrick J. Lynch (medical illustrator), Wikimedia Commons, CC BY 2.5. *http://commons.wikimedia.org/wiki/File:Human_skull_lateral_view.jpg*; and User:Kaligula, Wikimedia Commons,CC BY-SA 3.0. *http://commons.wikimedia.org/wiki/File:Human_dental_arches.svg*

LAB 27
p. 221: User:LadyofHats, Wikimedia Commons, Public domain. *http://commons.wikimedia.org/wiki/File%3AMain_protein_structure_levels_en.svg*

p. 222: University of California Museum of Paleontology's Understanding Evolution. *http://evolution.berkeley.edu/evolibrary/search/imagedetail.php?id=250&topic_id=&keywords=*